INFINITE FAMILIES OF EXACT SUMS OF SQUARES FORMULAS, JACOBI ELLIPTIC FUNCTIONS, CONTINUED FRACTIONS, AND SCHUR FUNCTIONS

By

Stephen C. Mil
Department of Mathe
The Ohio State University

Reprinted from a Special Issue of
The Ramanujan Journal
Volume 6, No. 1
March 2002

KLUWER ACADEMIC PUBLISHERS
Boston / Dordrecht / London

Developments in Mathematics

VOLUME 5

Series Editor:

Krishnaswami Alladi, *University of Florida, U.S.A.*

Aims and Scope

Developments in Mathematics is a book series publishing

(i) Proceedings of Conferences dealing with the latest research advances,

(ii) Research Monographs, and

(iii) Contributed Volumes focussing on certain areas of special interest.

Editors of conference proceedings are urged to include a few survey papers for wider appeal. Research monographs which could be used as texts or references for graduate level courses would also be suitable for the series. Contributed volumes are those where various authors either write papers or chapters in an organized volume devoted to a topic of special/current interest or importance. A contributed volume could deal with a classical topic which is once again in thc limelight owing to new developments.

THE RAMANUJAN JOURNAL

Volume 6, No. 1, March 2002

Preface .... *George E. Andrews* 5

Infinite Families of Exact Sums of Squares Formulas, Jacobi Elliptic Functions, Continued Fractions, and Schur Functions .... *Stephen C. Milne* 7

Distributors for North, Central and South America:
Kluwer Academic Publishers
101 Philip Drive
Assinippi Park
Norwell, Massachusetts 02061 USA
Telephone (781) 871-6600
Fax (781) 681-9045
E-Mail: kluwer@wkap.com

Distributors for all other countries:
Kluwer Academic Publishers Group
Post Office Box 322
3300 AH Dordrecht, THE NETHERLANDS
Telephone 31 786 576 000
Fax 31 786 576 474
E-Mail: services@wkap.nl

Electronic Services <http://www.wkap.nl>
ISBN 978-1-4419-5213-4

Library of Congress Cataloging-in-Publication Data

A C.I.P. Catalogue record for this book is available from the Library of Congress.

Printed on acid-free paper.

Printed in the United States of America.

THE RAMANUJAN JOURNAL, 6, 5–6, 2002

Preface

The study of sums of squares is one of the oldest and most revered in mathematics. In chapter 11 of his book, *Great Moments in Mathematics Before 1650*, Howard Eves says "Now in the history of mathematics there is one man who stands out as probably the first true genius in the field of number theory, and one whose works so profoundly influenced later European number theorists that the production of this work can well be labeled a GREAT MOMENT IN MATHEMATICS. The man is Diophantus of Alexandria, and the work alluded to is his famous *Arithmetica*."

This book posed numerous problems whose solutions were to be integers. Such problems have become known as Diophantine problems. Probably the most famous of all Diophantus' problems is Problem 8 of Book II of the *Arithmetica* which reads "To divide a given square number into two squares." As we are all aware, adjoining this problem in the margin of his copy of Bachet's translation, Fermat wrote his famous "Last Theorem" with the tantalizing comment that he "did indeed have a proof . . . but the margin is too narrow to contain it."

In Book IV of the *Arithmetica*, Diophantus considers problems involving sums of four squares, and Bachet asserted that Diophantus assumed that every number was either a square or the sum of two, three or four squares. Euler, Fermat and others attempted to prove this assertion, and in the middle of the eighteen century, Lagrange succeeded. From then until today discoveries about sums of squares have been of great significance in mathematics.

One of the milestones in the work on sums of squares was *Jacobi's Fundamenta Nova Theoriae Functionum Ellipticarum* in 1829. In this book, Jacobi developed the method of elliptic functions sufficiently to produce exact formulae for $r_s(n)$, the number of representations of n as a sum of s squares when $s = 2, 4, 6$ or 8.

At the beginning of the twentieth century, Mordell pioneered the application of the theory of modular forms to sums of squares problems. This approach was sufficiently powerful that it came to dominate much of the work in this area during the twentieth century.

In the mid-1980s, Grosswald published his book *Representations of Integers as the Sums of Squares*. In chapters 8 and 9 of this book, Grosswald illuminated the nineteenth century elliptic function methods of Jacobi, Glaisher and others. In chapter 10, he presented the modular form ideas of Mordell and others that flourished in the twentieth century.

Here at the beginning of the twenty-first century, we welcome Steve Milne's extensive contribution to the elliptic function tradition in the study of sums of squares. The genesis of Milne's studies lies in his decades-long development of generalized hypergeometric and q-hypergeometric series related to the classical groups. Subsequently he saw how to amalgamate those studies with his combinatorial insights and classical elliptic function theory to obtain the powerful results in this monograph. The original elliptic function methods provided formulas for $r_s(n)$ when s was small (usually ≤ 24). For example, examination of Jacobi's *Fundamenta Nova*, sections 40–42, reveals twenty plus identities related to generating functions for sums of squares. Milne produces infinite families of identities wherein the aforementioned identities of Jacobi are each the first member of one family. In addition, he proves the conjectured identities of Kac and Wakimoto, which involve triangular numbers and which arose in the study of Lie algebras. (It should be noted that Don Zagier has independently proved these conjectures utilizing the theory of modular forms.)

This impressive paper will undoubtedly spur others both in elliptic functions and in modular forms to build on these wonderful discoveries.

George E. Andrews

THE RAMANUJAN JOURNAL, 6, 7–149, 2002

Infinite Families of Exact Sums of Squares Formulas, Jacobi Elliptic Functions, Continued Fractions, and Schur Functions

STEPHEN C. MILNE* milne@math.ohio-state.edu
Department of Mathematics, The Ohio State University, Columbus, Ohio, 43210

Dedicated to the memory of Gian-Carlo Rota who encouraged me to write this paper in the present style

Received August 5, 2000; Accepted March 1, 2001

Abstract. In this paper we derive many infinite families of explicit exact formulas involving either squares or triangular numbers, two of which generalize Jacobi's 4 and 8 squares identities to $4n^2$ or $4n(n+1)$ squares, respectively, without using cusp forms. In fact, we similarly generalize to infinite families all of Jacobi's explicitly stated degree 2, 4, 6, 8 Lambert series expansions of classical theta functions. In addition, we extend Jacobi's special analysis of 2 squares, 2 triangles, 6 squares, 6 triangles to 12 squares, 12 triangles, 20 squares, 20 triangles, respectively. Our 24 squares identity leads to a different formula for Ramanujan's tau function $\tau(n)$, when n is odd. These results, depending on new expansions for powers of various products of classical theta functions, arise in the setting of Jacobi elliptic functions, associated continued fractions, regular C-fractions, Hankel or Turánian determinants, Fourier series, Lambert series, inclusion/exclusion, Laplace expansion formula for determinants, and Schur functions. The Schur function form of these infinite families of identities are analogous to the η-function identities of Macdonald. Moreover, the powers $4n(n+1)$, $2n^2+n$, $2n^2-n$ that appear in Macdonald's work also arise at appropriate places in our analysis. A special case of our general methods yields a proof of the two Kac–Wakimoto conjectured identities involving representing a positive integer by sums of $4n^2$ or $4n(n+1)$ triangular numbers, respectively. Our 16 and 24 squares identities were originally obtained via multiple basic hypergeometric series, Gustafson's C_ℓ nonterminating ${}_6\phi_5$ summation theorem, and Andrews' basic hypergeometric series proof of Jacobi's 2, 4, 6, and 8 squares identities. We have (elsewhere) applied symmetry and Schur function techniques to this original approach to prove the existence of similar infinite families of sums of squares identities for n^2 and $n(n+1)$ squares. Our sums of more than 8 squares identities are not the same as the formulas of Mathews (1895), Glaisher (1907), Sierpinski (1907), Uspensky (1913, 1925, 1928), Bulygin (1914, 1915), Ramanujan (1916), Mordell (1917, 1919), Hardy (1918, 1920), Bell (1919), Estermann (1936), Rankin (1945, 1962), Lomadze (1948), Walton (1949), Walfisz (1952), Ananda-Rau (1954), van der Pol (1954), Krätzel (1961, 1962), Bhaskaran (1969), Gundlach (1978), Kac and Wakimoto (1994), and, Liu (2001). We list these authors by the years their work appeared.

Key words: Jacobi elliptic functions, associated continued fractions, regular C-fractions, Hankel or Turánian determinants, Fourier series, Lambert series, Eisenstein series, inclusion/exclusion, Laplace expansion formula for determinants, Schur functions, multiple basic hypergeometric series, C_ℓ nonterminating ${}_6\phi_5$ summation theorem, lattice sums

2000 Mathematics Subject Classification: Primary—11E25, 33E05; Secondary—05A15, 33D70

*S.C. Milne was partially supported by National Security Agency grants MDA 904-93-H-3032, MDA 904-97-1-0019 and MDA 904-99-1-0003.

1. Introduction

In this paper we derive many infinite families of explicit exact formulas involving either squares or triangular numbers, two of which generalize Jacobi's [117] 4 and 8 squares identities to $4n^2$ or $4n(n+1)$ squares, respectively, without using cusp forms. In fact, we similarly generalize to infinite families all of Jacobi's [117] explicitly stated degree 2, 4, 6, 8 Lambert series expansions of classical theta functions. In addition, we extend Jacobi's special analysis of 2 squares, 2 triangles, 6 squares, 6 triangles to 12 squares, 12 triangles, 20 squares, 20 triangles, respectively. We also utilize a special case of our general analysis, outlined in [162], to prove the two Kac–Wakimoto [120, p. 452] conjectured identities involving representing a positive integer by sums of $4n^2$ or $4n(n+1)$ triangular numbers, respectively. Zagier in [254] has also recently independently proven these two identities. In addition, he proved the more general Conjecture 7.2 of Kac–Wakimoto [120, p. 451]. The $n = 1$ cases of these two Kac–Wakimoto conjectured identities for triangular numbers are equivalent to the classical identities of Legendre [140], [21, Eqs. (ii) and (iii), p. 139]. Recently, in [183], Ono utilized an elementary modular forms involution to transform Zagier's proof and/or formulation in [254] of Corollary 7.6 into elegant corresponding sums of squares formulas for $4n^2$ or $4n(n+1)$ squares, respectively. These formulas of Ono are different than those in Theorems 5.3–5.6, and Theorems 7.1 and 7.2. Our 24 squares identity leads to a different formula for Ramanujan's [194] tau function $\tau(n)$, when n is odd.

All of this work depends upon new expansions for powers of various products of the classical theta functions

$$\vartheta_3(0,q) := \sum_{j=-\infty}^{\infty} q^{j^2}, \tag{1.1}$$

$$\vartheta_2(0,q) := \sum_{j=-\infty}^{\infty} q^{(j+1/2)^2}, \tag{1.2}$$

and $\vartheta_4(0,q) := \vartheta_3(0,-q)$, where $\vartheta_3(0,q)$, $\vartheta_2(0,q)$, and $\vartheta_4(0,q)$ are the $z = 0$ cases of the theta functions $\vartheta_3(z,q)$, $\vartheta_2(z,q)$, and $\vartheta_4(z,q)$ in [250, pp. 463–464]. We first give a single determinant form of these expansions, then (where applicable) a sum of determinants form, and finally a Schur function form. The single determinant form of our identities includes expressing the quadratic powers $\vartheta_3(0,-q)^{4n^2}$, $\vartheta_3(0,-q)^{4n(n+1)}$, $\vartheta_2(0,q)^{4n^2}$, and $\vartheta_2(0,q^{1/2})^{4n(n+1)}$ of classical theta functions as a constant multiple of an $n \times n$ Hankel determinant whose entries are either certain Lambert series, or Lambert series plus nonzero constants, for ϑ_2 and ϑ_3, respectively. Our sums of squares identities arise from the sum of determinants form of our expansions by writing $\vartheta_3(0,-q)^{4n^2}$ and $\vartheta_3(0,-q)^{4n(n+1)}$ as explicit polynomials of degree n in $2n-1$ Lambert series similar to those in Jacobi's 4 and 8 squares identities. The Schur function form of our analogous identities for $\vartheta_2(0,q)^{4n^2}$ and $\vartheta_2(0,q^{1/2})^{4n(n+1)}$ completes our proof of the Kac–Wakimoto conjectured identities for triangular numbers in [120, p. 452]. Depending on the analysis, we either use $\vartheta_3(0,-q)$ or $\vartheta_4(0,q)$ in many of our identities. Some of the above work has already been announced in [162]. We present a number of additional related results in [164–166].

Our derivation of the above infinite families of identities entails the analysis and combinatorics of Jacobi elliptic functions, associated continued fractions, regular C-fractions, Hankel or Turánian determinants, Fourier series, Lambert series, inclusion/exclusion, the Laplace expansion formula for determinants, and Schur functions. This background material is contained in [2, 15, 16, 21, 22, 43, 63, 79, 92, 93, 103, 104, 111, 119, 125, 134, 149, 153, 185, 204, 207, 215, 248, 250, 259]. In order to make the connection with divisor sums, we follow Jacobi in emphasizing Lambert series $\sum_{r=1}^{\infty} f(r)q^r/(1-q^r)$, as defined in [8, Example 14, p. 24], as much as possible. (For Glaisher's historical note of where these series first appeared in Lambert's writings see [84, Section 13, p. 163].) At the end of Section 2 we point out the relationship between the Lambert series $U_{2m-1}(q)$, $G_{2m+1}(q)$, $C_{2m-1}(q)$, $D_{2m+1}(q)$ appearing in our sums of squares and sums of triangles identities and the Fourier expansions of the classical Eisenstein series $E_n(\tau)$, with $q := \exp(2\pi i\tau)$ and n an even positive integer, as given by [20, p. 318] and [204, pp. 194–195]. For the convenience of the reader we often reference or write down the $n=1$ and $n=2$ cases of our general infinite families of identities. These special cases can also be given short modular forms verification proofs.

The problem of representing an integer as a sum of squares of integers is one of the chestnuts of number theory, where new significant contributions occur infrequently over the centuries. The long and interesting history of this topic is surveyed in [11, 12, 26, 43, 94, 101, 102, 189, 201, 205, 230, 242] and chapters 6–9 of [56]. The review article [223] presents many questions connected with representations of integers as sums of squares. Direct applications of sums of squares to lattice point problems and crystallography models in theoretical chemistry can be found in [28, 90]. One such example is the computation of the constant Z_N that occurs in the evaluation of a certain Epstein zeta function, needed in the study of the stability of rare gas crystals, and in that of the so-called *Madelung constants* of ionic salts. More theoretical applications to "theta series" appear in [51].

The s squares problem is to count the number $r_s(n)$ of integer solutions $(x_1, \ldots, x_s)$ of the Diophantine equation

$$x_1^2 + \cdots + x_s^2 = n, \tag{1.3}$$

in which changing the sign or order of the x_i's give distinct solutions.

Diophantus (325–409 A.D.) knew that no integer of the form $4n-1$ is a sum of two squares. Girard conjectured in 1632 that n is a sum of two squares if and only if all prime divisors q of n with $q \equiv 3 \pmod 4$ occur in n to an even power. Fermat in 1641 gave an "irrefutable proof" of this conjecture. Euler gave the first known proof in 1749. Early explicit formulas for $r_2(n)$ were given by Legendre in 1798 and Gauß in 1801. It appears that Diophantus was aware that all positive integers are sums of four integral squares. Bachet conjectured this result in 1621, and Lagrange gave the first proof in 1770.

Jacobi in his famous *Fundamenta Nova* [117] of 1829 introduced elliptic and theta functions, and utilized them as tools in the study of (1.3). Motivated by Euler's work on 4 squares, Jacobi observed that the number $r_s(n)$ of integer solutions of (1.3) is also determined by

$$\vartheta_3(0,q)^s := 1 + \sum_{n=1}^{\infty} r_s(n)q^n, \tag{1.4}$$

where $\vartheta_3(0,q)$ is the classical theta function in (1.1).

Jacobi then used his theory of elliptic and theta functions to derive remarkable identities for the $s=2,4,6,8$ cases of $\vartheta_3(0,q)^s$. He immediately obtained elegant explicit formulas for $r_s(n)$, where $s=2,4,6,8$. We find it more convenient to work with Jacobi's equivalent identities for the $s=2,4,6,8$ cases of $\vartheta_3(0,-q)^s$. Dealing with powers of $\vartheta_3(0,-q)$ simplifies somewhat many of our identities in Sections 5, 7, and 8, especially the multiple power series of Section 7. The signs and infinite products in the classical formulas for $\vartheta_3(0,-q)^{16}$ and $\vartheta_3(0,-q)^{24}$ in Theorems 1.2 and 1.3 below are simpler. Finally, $\vartheta_3(0,-q)^s$ arises more naturally in the basic hypergeometric series or multiple basic hypergeometric series analysis in [4] and [6, pp. 506–508], or [167], respectively.

We recall Jacobi's identities from [117, Eq. (10), Section 40; Eq. (36), Section 40] and [117, Eq. (7), Section 42] for the $s=4$ and 8 cases of $\vartheta_3(0,-q)^s$ in the following theorem.

Theorem 1.1 (Jacobi).

$$\vartheta_3(0,-q)^4 = 1-8\sum_{r=1}^{\infty}(-1)^{r-1}\frac{rq^r}{1+q^r} = 1+8\sum_{n=1}^{\infty}(-1)^n\left[\sum_{\substack{d|n,d>0\\ 4\nmid d}} d\right]q^n, \tag{1.5}$$

and

$$\vartheta_3(0,-q)^8 = 1+16\sum_{r=1}^{\infty}(-1)^r\frac{r^3q^r}{1-q^r} = 1+16\sum_{n=1}^{\infty}\left[\sum_{d|n,d>0}(-1)^d d^3\right]q^n. \tag{1.6}$$

Consequently, we have respectively,

$$r_4(n) = 8\sum_{\substack{d|n,d>0\\ 4\nmid d}} d \quad \text{and} \quad r_8(n) = 16\sum_{d|n,d>0}(-1)^{n+d}d^3. \tag{1.7}$$

Note that Jacobi's identities for the $s=4$ and 8 cases of $\vartheta_3(0,q)^s$ appear in [117, Eq. (8), Section 40; Eq. (34), Section 40] and [117, Eq. (8), Section 42], respectively.

In general it is true that

$$r_{2s}(n) = \delta_{2s}(n) + e_{2s}(n), \tag{1.8}$$

where $\delta_{2s}(n)$ is a divisor function and $e_{2s}(n)$ is a function of order substantially lower than that of $\delta_{2s}(n)$. If $2s=2,4,6,8$, then $e_{2s}(n)=0$, and (1.8) becomes Jacobi's formulas for $r_{2s}(n)$, including (1.7). On the other hand [203, 204], if $2s>8$ then $e_{2s}(n)$ is never 0. The function $e_{2s}(n)$ is the coefficient of q^n in a suitable "cusp form". The difficulties of computing (1.8), and especially the "nondominate" term $e_{2s}(n)$, increase rapidly with $2s$. The modular function approach to (1.8) and the cusp form $e_{2s}(n)$ is discussed in [123, 203]; [204, pp. 241–244]. For $2s>8$ modular function methods such as those in [95, 99, 100, 147, 172, 202], or the more classical elliptic function approach of [14, 25, 33, 34, 39, 126, 127], [155, pp. 140–143], [242, pp. 204–211], are used to determine general formulas for $\delta_{2s}(n)$ and $e_{2s}(n)$ in (1.8). Explicit, exact examples of (1.8) have been worked out for $2 \leq 2s \leq 32$. Similarly, explicit formulas for $r_s(n)$ have been found for (odd) $s < 32$. Alternate, elementary approaches to sums of squares formulas can be found in [154, 212, 227–230].

We next consider classical analogs of (1.5) and (1.6) corresponding to the $s = 8$ and 12 cases of (1.8).

Glaisher [87, p. 210] utilized elliptic function methods, rather than modular functions, to prove

Theorem 1.2 (Glaisher).

$$\vartheta_3(0,-q)^{16} = 1 + \tfrac{32}{17} \sum_{y_1, m_1 \geq 1} (-1)^{m_1} m_1^7 q^{m_1 y_1} \tag{1.9a}$$

$$- \tfrac{512}{17} q(q;q)_\infty^8 (q^2;q^2)_\infty^8 \tag{1.9b}$$

where we have

$$(q;q)_\infty := \prod_{r \geq 1} (1 - q^r), \quad \textit{with } 0 < |q| < 1. \tag{1.10}$$

Glaisher took the coefficient of q^n to obtain $r_{16}(n)$. The same formula appears in [204, Eq. (7.4.32), p. 242].

In order to find $r_{24}(n)$, Ramanujan [194, Entry 7, Table VI], see also [204, Eq. (7.4.37), p. 243] and [89], first proved

Theorem 1.3 (Ramanujan). *Let* $(q;q)_\infty$ *be defined by* (1.10). *Then*

$$\vartheta_3(0,-q)^{24} = 1 + \tfrac{16}{691} \sum_{y_1, m_1 \geq 1} (-1)^{m_1} m_1^{11} q^{m_1 y_1} \tag{1.11a}$$

$$- \tfrac{33152}{691} q(q;q)_\infty^{24} - \tfrac{65536}{691} q^2 (q^2;q^2)_\infty^{24}. \tag{1.11b}$$

An analysis of (1.11b) depends upon Ramanujan's [194] tau function $\tau(n)$ defined by

$$q(q;q)_\infty^{24} := \sum_{n=1}^{\infty} \tau(n) q^n. \tag{1.12}$$

For example, $\tau(1) = 1$, $\tau(2) = -24$, $\tau(3) = 252$, $\tau(4) = -1472$, $\tau(5) = 4830$, $\tau(6) = -6048$, and $\tau(7) = -16744$. Ramanujan [194, Eq. (103)] conjectured, and Mordell [171] proved that $\tau(n)$ is multiplicative.

Taking the coefficient of q^n in (1.11) yields the classical formula [107, p. 216], [194], [204, Eq. (7.4.37), p. 243] for $r_{24}(n)$ given by

$$r_{24}(n) = \tfrac{16}{691} (-1)^n \sigma_{11}^\dagger(n) + \tfrac{128}{691} \{(-1)^{(n-1)} 259 \tau(n) - 512 \tau(\tfrac{n}{2})\}, \tag{1.13}$$

where

$$\sigma_r^{\dagger}(n) := \sum_{d|n, d>0} (-1)^d d^r, \tag{1.14}$$

and $\tau(x) = 0$ if x is not an integer.

The classical formula (1.13) can be rewritten in terms of divisor functions by appealing to the formula for $\tau(n)$ from [130, Eq. (24), p. 34; and, Eq. (11.1), p. 36], [140, Eq. (9), p. 111], [8, Example 10, p. 140] given by

$$\tau(n) = \tfrac{65}{756}\sigma_{11}(n) + \tfrac{691}{756}\sigma_5(n) - \tfrac{691}{3}\sum_{m=1}^{n-1} \sigma_5(m)\sigma_5(n-m), \tag{1.15}$$

where

$$\sigma_r(n) := \sum_{d|n, d>0} d^r. \tag{1.16}$$

Another useful formula for $\tau(n)$ is [140, Eq. (10), p. 111]. Many similar identities attributed to Ramanujan appear in [128–132], and an exposition of classical results on $\tau(n)$ can be found in [27].

For convenience, we work with (1.15) in the form

$$q(q;q)_\infty^{24} = \tfrac{65}{756} V_{11}(q) + \tfrac{691}{756} V_5(q) - \tfrac{691}{3} V_5^2(q), \tag{1.17}$$

where

$$V_s \equiv V_s(q) := \sum_{r=1}^{\infty} \frac{r^s q^r}{1-q^r} = \sum_{n=1}^{\infty} \left[\sum_{d|n, d>0} d^s \right] q^n = \sum_{y_1, m_1 \geq 1} m_1^s q^{m_1 y_1}. \tag{1.18}$$

The generating function version of applying (1.15) to (1.13) now becomes

Theorem 1.4. *Let $G_s(q)$ and $V_s(q)$ be defined by* (1.23) *and* (1.18), *respectively. Then,*

$$\vartheta_3(0,-q)^{24} = 1 + \tfrac{16}{691} G_{11}(q) - \tfrac{538720}{130599} V_{11}(q) - \tfrac{1064960}{130599} V_{11}(q^2) - \tfrac{8288}{189} V_5(q) - \tfrac{16384}{189} V_5(q^2) + \tfrac{33152}{3} V_5^2(q) + \tfrac{65536}{3} V_5^2(q^2). \tag{1.19}$$

If all we wanted to do is write $\vartheta_3(0,-q)^{24}$ as a sum of products of at most 6 or 3 Lambert series, we just take either the 6th or 3rd power of both sides of (1.5) or (1.6), respectively. A slightly more interesting expansion of $\vartheta_3(0,-q)^{24}$ as a sum of products of at most 3 Lambert series results from applying the formula for $\tau(n)$ in [8, Example 13, p. 24] to (1.11)–(1.14). Equation (1.19) which expresses $\vartheta_3(0,-q)^{24}$ as a sum of products of at most two Lambert series lies deeper.

One of the main motivations for this paper was to generalize Theorem 1.1 to $4n^2$ or $4n(n+1)$ squares, respectively, without using cusp forms such as (1.9b) and (1.11b), while still utilizing just polynomials of degree n in $2n-1$ Lambert series similar to either (1.5) or (1.6), respectively. This condition on the maximal number n of Lambert series factors in each term is consistent with the η-function identities in Appendix I of Macdonald [151], the two Kac–Wakimoto conjectured identities for triangular numbers in [120, p. 452], and the above discussion of (1.19). Essentially, expansions of $\vartheta_3(0,-q)^N$, for N large, into a polynomial of at most degree M in Lambert series is "trivial" if $M=\alpha N$, and lies deeper if $M=\alpha\sqrt{N}$, where $\alpha>0$ is a constant.

We carry out the above program in Theorems 5.4 and 5.6 below. Here, we state the $n=2$ cases, which determine different formulas for 16 and 24 squares.

Theorem 1.5.

$$\vartheta_3(0,-q)^{16} = 1-\tfrac{32}{3}\,(U_1+U_3+U_5)+\tfrac{256}{3}\left(U_1U_5-U_3^2\right), \tag{1.20}$$

where

$$\begin{aligned} U_s \equiv U_s(q) &:= \sum_{r=1}^{\infty}(-1)^{r-1}\frac{r^sq^r}{1+q^r} = \sum_{n=1}^{\infty}\Bigg[\sum_{d|n,d>0}(-1)^{d+n/d}d^s\Bigg]q^n \\ &= \sum_{y_1,m_1\geq 1}(-1)^{y_1+m_1}m_1^sq^{m_1y_1}, \end{aligned} \tag{1.21}$$

and

$$\vartheta_3(0,-q)^{24} = 1+\tfrac{16}{9}\,(17G_3+8G_5+2G_7)+\tfrac{512}{9}\left(G_3G_7-G_5^2\right), \tag{1.22}$$

where

$$\begin{aligned} G_s \equiv G_s(q) &:= \sum_{r=1}^{\infty}(-1)^{r}\frac{r^sq^r}{1-q^r} = \sum_{n=1}^{\infty}\Bigg[\sum_{d|n,d>0}(-1)^{d}d^s\Bigg]q^n \\ &= \sum_{y_1,m_1\geq 1}(-1)^{m_1}m_1^sq^{m_1y_1}. \end{aligned} \tag{1.23}$$

A more compact way of writing Theorem 1.5 as a single determinant is provided by

Theorem 1.6.

$$\vartheta_3(0,-q)^{16} = \tfrac{1}{3}\det\begin{vmatrix}16U_1-2 & 16U_3+1\\ 16U_3+1 & 16U_5-2\end{vmatrix}, \tag{1.24}$$

and

$$\vartheta_3(0,-q)^{24} = \tfrac{1}{3^2}\det\begin{vmatrix}16G_3+1 & 16G_5-2\\ 32G_5-4 & 32G_7+17\end{vmatrix}, \tag{1.25}$$

where $U_s\equiv U_s(q)$ and $G_s\equiv G_s(q)$ are given by (1.21) *and* (1.23), *respectively.*

The general case of Theorem 1.6 for $\vartheta_3(0,-q)^{4n^2}$ and $\vartheta_3(0,-q)^{4n(n+1)}$ is given by Theorems 5.3 and 5.5.

The formulas for $r_{16}(n)$ and $r_{24}(n)$ corresponding to Theorem 1.5 are given by

Theorem 1.7. *Let n be any positive integer. Then*

$$r_{16}(n) = -(-1)^n \cdot \tfrac{32}{3}[\tilde{\sigma_1}(n) + \tilde{\sigma_3}(n) + \tilde{\sigma_5}(n)] \tag{1.26a}$$

$$+(-1)^n \cdot \tfrac{256}{3}\sum_{m=1}^{n-1}[\tilde{\sigma_1}(m)\tilde{\sigma_5}(n-m) - \tilde{\sigma_3}(m)\tilde{\sigma_3}(n-m)], \tag{1.26b}$$

where

$$\tilde{\sigma_r}(n) := \sum_{d|n,d>0} (-1)^{d+n/d} d^r, \tag{1.27}$$

and

$$r_{24}(n) = (-1)^n \cdot \tfrac{16}{9}[17 \cdot \sigma_3^\dagger(n) + 8 \cdot \sigma_5^\dagger(n) + 2 \cdot \sigma_7^\dagger(n)] \tag{1.28a}$$

$$+(-1)^n \cdot \tfrac{512}{9}\sum_{m=1}^{n-1}[\sigma_3^\dagger(m)\sigma_7^\dagger(n-m) - \sigma_5^\dagger(m)\sigma_5^\dagger(n-m)], \tag{1.28b}$$

where $\sigma_r^\dagger(n)$ is defined by (1.14).

The elementary analysis in [94, pp. 122–123, 125] applied to (1.26) and (1.28) immediately implies that the dominate terms (1.26b) and (1.28b) for $r_{16}(n)$ and $r_{24}(n)$ have orders of magnitude n^7 and n^{11}, respectively. Furthermore, the "remainder terms" (1.26a) and (1.28a) have lower orders of magnitude n^5 and n^7, respectively. The dominate term estimates are consistent with [94, Eq. (9.20), p. 122]. We provide a more detailed analyis of these estimates for (1.26) and (1.28) at the end of Section 5.

In Sections 5 and 7 we present the infinite families of explicit exact formulas that generalize Theorems 1.1, 1.5, and 1.6.

In the case where n is an odd integer [in particular an odd prime], equating (1.11) and (1.22) yields two formulas for $\tau(n)$, which we have presented in [162]. The first one is somewhat similar to (1.15), and both are different from Dyson's [61] formula. We first obtain

Theorem 1.8. *Let $\tau(n)$ be defined by* (1.12) *and let n be odd. Then*

$$259\tau(n) = \tfrac{1}{2^3 \cdot 3^2}[17 \cdot 691\sigma_3(n) + 8 \cdot 691\sigma_5(n) + 2 \cdot 691\sigma_7(n) - 9\sigma_{11}(n)]$$

$$- \tfrac{691 \cdot 2^2}{3^2}\sum_{m=1}^{n-1}[\sigma_3^\dagger(m)\sigma_7^\dagger(n-m) - \sigma_5^\dagger(m)\sigma_5^\dagger(n-m)], \tag{1.29}$$

where $\sigma_r(n)$ and $\sigma_r^\dagger(n)$ are defined by (1.16) *and* (1.14), *respectively.*

Remark. We can use (1.29) to compute $\tau(n)$ in $\leq 6n \ln n$ steps when n is an odd integer, and (1.15) to compute $\tau(n)$ in $\leq 3n \ln n$ steps when n is any positive integer. On the other hand, this may also be done in $n^{2+\epsilon}$ steps by appealing to Euler's infinite-product-representation algorithm (EIPRA) [5, p. 104] applied to $(q;q)_\infty^{24}$ in (1.12).

A different simplification involving a power series formulation of (1.22) leads to

Theorem 1.9. *Let $\tau(n)$ be defined by* (1.12) *and let $n \geq 3$ be odd. Then*

$$259\tau(n) = -\tfrac{1}{2^3}\sum_{d|n,d>0} d^{11} + \tfrac{691}{2^3 \cdot 3^2}\sum_{d|n,d>0} d^3(17+8d^2+2d^4) \tag{1.30a}$$

$$-\tfrac{691\cdot 2^2}{3^2}\sum_{\substack{m_1>m_2\geq 1\\ m_1+m_2\leq n\\ gcd(m_1,m_2)|n}} (-1)^{m_1+m_2}(m_1m_2)^3(m_1^2-m_2^2)^2 \sum_{\substack{y_1,y_2\geq 1\\ m_1y_1+m_2y_2=n}} 1. \tag{1.30b}$$

Remark. The inner sum in (1.30b) counts the number of solutions (y_1, y_2) of the classical linear Diophantine equation $m_1y_1 + m_2y_2 = n$. This relates (1.30) to the combinatorics in Sections 4.6 and 4.7 of [215].

Equation (1.15) and Eq. (11.3) of [130, p. 36] in the form

$$\tau(n) = \tfrac{691}{1800}\sigma_3(n) + \tfrac{691}{900}\sigma_7(n) - \tfrac{91}{600}\sigma_{11}(n) + \tfrac{2764}{15}\sum_{m=1}^{n-1}\sigma_3(m)\sigma_7(n-m), \tag{1.31}$$

yield formulas for $\tau(n)$ that are similar to (1.29) and (1.30). That is, motivated by Theorems 1.8 and 1.9, a linear combination of (1.15) and (1.31) leads to

Theorem 1.10. *Let n be any positive integer, and let $\tau(n)$ and $\sigma_r(n)$ be defined by* (1.12) *and* (1.16), *respectively. Then*

$$\tau(n) = \tfrac{1}{2^3\cdot 3^3\cdot 5\cdot 7}[3\cdot 7\cdot 691\sigma_3(n) + 2^3\cdot 5\cdot 691\sigma_5(n) + 2\cdot 3\cdot 7\cdot 691\sigma_7(n) - 13\cdot 241\sigma_{11}(n)]$$

$$+\tfrac{691\cdot 2^2}{3^3}\sum_{m=1}^{n-1}[\sigma_3(m)\sigma_7(n-m) - \sigma_5(m)\sigma_5(n-m)], \tag{1.32}$$

and

$$\tau(n) = \tfrac{1}{2^3\cdot 3^3\cdot 5\cdot 7}\sum_{d|n,d>0} d^3(3\cdot 7\cdot 691 + 2^3\cdot 5\cdot 691d^2 + 2\cdot 3\cdot 7\cdot 691d^4 - 13\cdot 241d^8)$$

$$+\tfrac{691\cdot 2^2}{3^3}\sum_{\substack{m_1>m_2\geq 1\\ m_1+m_2\leq n\\ gcd(m_1,m_2)|n}} (m_1m_2)^3(m_1^2-m_2^2)^2 \sum_{\substack{y_1,y_2\geq 1\\ m_1y_1+m_2y_2=n}} 1. \tag{1.33}$$

Equating (1.29) and (1.32) immediately gives

Corollary 1.11. *Let n be odd, and let $\sigma_r^\dagger(n)$ and $\sigma_r(n)$ be defined by* (1.14) *and* (1.16), *respectively. Then*

$$2160\sum_{m=1}^{n-1}[\sigma_3^\dagger(m)\sigma_7^\dagger(n-m)-\sigma_5^\dagger(m)\sigma_5^\dagger(n-m)]+186480\sum_{m=1}^{n-1}$$
$$\times[\sigma_3(m)\sigma_7(n-m)-\sigma_5(m)\sigma_5(n-m)]$$
$$=759\sigma_3(n)-200\sigma_5(n)-642\sigma_7(n)+83\sigma_{11}(n). \tag{1.34}$$

We next give a brief description of some of the essential ingredients in our derivation of the sums of squares and related identities in this paper.

The first is a classical result of Heilermann [103, 104], more recently presented in [119, Theorem 7.14, pp. 244–246], in which Hankel determinants whose entries are the coefficients in a formal power series L can be expressed as a certain product of the "numerator" coefficients of the associated continued fraction J corresponding to L, provided J exists. A similar result holds for the related χ determinants. We apply Heilermann's product formulas for both $n \times n$ Hankel and the related χ determinants to Rogers' [207], Stieltjes' [218–220], and Ismail and Masson's [111] associated continued fraction and/or regular C-fraction expansions of the Laplace transform of a small number of Jacobi elliptic functions such as $\mathrm{sn}(u,k)$, $\mathrm{cn}(u,k)$, $\mathrm{dn}(u,k)$, $\mathrm{sn}^2(u,k)$, $\mathrm{sn}(u,k)\,\mathrm{cn}(u,k)/\mathrm{dn}(u,k)$, and $\mathrm{sn}(u,k)\,\mathrm{cn}(u,k)$. Modular transformations, row and column operations, and Heilermann's [103, 104], [119, Theorem 7.14, pp. 244–246; Theorem 7.2, pp. 223–224] correspondence theorems between formal power series and both types of continued fractions enables us to carry out a similar product formula analysis for our associated continued fraction and/or regular C-fraction expansions of the *formal* Laplace transform of a number of ratios of Jacobi elliptic functions in which $\mathrm{cn}(u,k)$ is in the denominator. These include both $\mathrm{sn}(u,k)\,\mathrm{dn}(u,k)/\mathrm{cn}(u,k)$ and $\mathrm{sn}^2(u,k)\,\mathrm{dn}^2(u,k)/\mathrm{cn}^2(u,k)$. Al-Salam and Carlitz [2, pp. 97–99] have already applied Heilermann's product formulas to Rogers' and Stieltjes' continued fraction expansions of the Laplace transform of $\mathrm{sn}(u,k)$, $\mathrm{cn}(u,k)$, $\mathrm{dn}(u,k)$, and $\mathrm{sn}^2(u,k)$ to obtain the product formulas for the corresponding Hankel determinants. The direct relationship between orthogonal polynomials, J-fraction expansions, and some of these continued fraction expansions of the Laplace transform of Jacobi elliptic functions is surveyed in [19, 38, 44, 46, 47, 49, 50, 69, 92, 109, 110, 115, 116, 119, 148, 231–235].

All of the above analysis produces a large number of product formulas for $n \times n$ Hankel or related χ determinants whose entries (polynomials in k^2) are the coefficients in the formal Laplace transform of the Maclaurin series expansion (about $u=0$) of certain ratios $f_1(u,k)$ of Jacobi elliptic functions, with k the modulus. These formulas express our determinants as a constant multiple of a simple polynomial in k^2. By an analysis similar to that in [15, 16, 21, 22, 38, 259] we compare the Maclaurin series expansion of $f_1(u,k)$ to its Fourier series. This immediately leads to a formula which factors the entries in our determinants into the product of a Lambert series plus a constant, a simple function of k, and a suitable negative integral power of the Jacobi elliptic function parameter $z := {}_2F_1(1/2, 1/2; 1; k^2) = 2\boldsymbol{K}(k)/\pi \equiv 2\boldsymbol{K}/\pi$, with $\boldsymbol{K}(k) \equiv \boldsymbol{K}$ the complete elliptic integral of the first kind in [134, Eq. (3.1.3), p. 51], and k the modulus. We next solve for the resulting power of z in our product formula for the $n \times n$ Hankel or related χ determinants. Appealing to well-known equalities such as

$z = \vartheta_3(0,q)^2$ and $zk = \vartheta_2(0,q)^2$, where $q := \exp(-\pi K(\sqrt{1-k^2})/K(k))$, and simplifying, we are able to establish the single determinant form of our identities. Our proofs of the single determinant form of the sums of squares identities in Theorems 5.3 and 5.5 utilizes the ratios of Jacobi elliptic functions $f_1(u,k) := \mathrm{sc}(u,k)\,\mathrm{dn}(u,k)$ and $f_1(u,k) := \mathrm{sc}^2(u,k)\,\mathrm{dn}^2(u,k)$, respectively. Similarly, the corresponding analysis for Theorem 5.11 in our proof of the Kac–Wakimoto conjectures utilizes $f_1(u,k) := \mathrm{sd}(u,k)\,\mathrm{cn}(u,k)$ and $f_1(u,k) := \mathrm{sn}^2(u,k)$.

An inclusion/exclusion argument applied to the above single determinant form of our identities yields a sum of determinants formulation. We transform these latter identities into a multiple power series Schur function form by employing classical properties of Schur functions [153], symmetry and skew-symmetry arguments, row and column operations, and the Laplace expansion formula [118, pp. 396–397] for a determinant.

We organize our paper as follows. In Section 2 we compare Fourier and Maclaurin series expansions of various ratios $f_1(u,k)$ of Jacobi elliptic functions to derive the Lambert series formulas we need. Section 3 contains a summary of the associated continued fraction and regular C-fraction expansions of the Laplace transform of these ratios $f_1(u,k)$. This includes a sketch of Rogers' [207] integration-by-parts derivation of several of these expansions, followed by his application of Landen's transformation [134, Eq. (3.9.15), (3.9.16), (3.9.17), pp. 78–79] to one of them. Here, we have continued fraction expansions of both Laplace transforms and formal Laplace transforms. Section 4 utilizes Theorems 7.14 and 7.2 of [119, pp. 244–246; pp. 223–224], row and column operations, and modular transformations to deduce our Hankel and χ determinant evaluations from the continued fraction expansions of Section 3. In Section 5 we first establish an inclusion-exclusion lemma, recall necessary elliptic function parameter relations, and then obtain the single determinant and sum of determinants form of our infinite families of sums of squares and related identities. We also obtain (see Theorem 5.19) our generalization to infinite families all 21 of Jacobi's [117, Sections 40–42] explicitly stated degree 2, 4, 6, 8 Lambert series expansions of classical theta functions. An elegant generating function involving the number of ways of writing N as a sum of $2n$ squares and $(2n)^2$ triangular numbers appears in Corollary 5.15. Section 6 contains the derivation of two key theorems which expand certain general $n \times n$ determinants, whose entries are either constants or Lambert series, into a multiple power series whose terms include classical Schur functions [153] as factors. Section 7 applies the key determinant expansion formulas in Section 6 to most of the main identities in Section 5 to obtain the Schur function form of our infinite families of sums of squares and related identities. These include the two Kac–Wakimoto [120, p. 452] conjectured identities involving representing a positive integer by sums of $4n^2$ or $4n(n+1)$ triangular numbers, respectively. The $n = 1$ case is equivalent to the classical identities of Legendre [139], [21, Eqs. (ii) and (iii), p. 139]. In addition, we obtain our analog of these Kac–Wakimoto identities which involves representing a positive integer by sums of $2n$ squares and $(2n)^2$ triangular numbers. Motivated by the analysis in [41, 42], we use Jacobi's transformation of the theta functions ϑ_4 and ϑ_2 to derive a direct connection between our identities involving $4n^2$ or $4n(n+1)$ squares, and the identities involving $4n^2$ or $4n(n+1)$ triangular numbers. In a different direction we also apply the classical techniques in [90, pp. 96–101] and [28, pp. 288–305] to several of the theorems in this section to obtain the corresponding infinite families of lattice sum transformations. Our explicit multiple power series formulas for 16, 24, 36, and 48 squares appear in Section 8.

The Schur function form of our infinite families of identities are analogous to the η-function identities in Appendix I of Macdonald [151]. Moreover, the powers $4n(n+1)$, $2n^2+n$, $2n^2-n$ that appear in Macdonald [151] also arise at appropriate places in our analysis. An important part of our approach to the infinite families of identities in this paper is based upon a limiting process (computing associated continued fraction expansions), followed by a sieving procedure (inclusion/exclusion). On the other hand, the derivation of the η-function and related identities in [76, 77, 141, 142, 151] relies on first a sieving procedure, followed by taking limits.

Theorems 1.5 and 1.6 were originally obtained via multiple basic hypergeometric series [143, 158–161, 163, 168, 169] and Gustafson's [96] C_ℓ nonterminating ${}_6\phi_5$ summation theorem combined with Andrews' [4], [6, pp. 506–508], [79, pp. 223–226] basic hypergeometric series proof of Jacobi's 2, 4, 6, and 8 squares identities, and computer algebra [251]. We have in [167] applied symmetry and Schur function techniques to this original approach to prove the existence of similar infinite families of sums of squares identities for n^2 or $n(n+1)$ squares, respectively.

Our sums of more than 8 squares identities are not the same as the formulas of Mathews [154], Glaisher [85–88], Sierpinski [212], Uspensky [227–229], Bulygin [33, 34], Ramanujan [194], Mordell [172, 173], Hardy [99, 100], Bell [14], Estermann [64], Rankin [201, 202], Lomadze [147], Walton [249], Walfisz [247], Ananda-Rau [3], van der Pol [187], Krätzel [126, 127], Bhaskaran [25], Gundlach [95], Kac and Wakimoto [120], and Liu [146].

We have found in [164–166] a number of additional new results involving or inspired by Hankel determinants. In [166] we apply the Hankel determinant evaluations in the present paper to the analysis in [119, pp. 244–250] to yield a large number of more complex χ determinant evaluations. All of these determinant evaluations lead to new determinantal formulas for powers of classical theta functions. The paper [164] presents some new evaluations of Hankel determinants of Eisenstein series which generalize the classical formula for the modular discriminant Δ in [20, Entry 12(i), p. 326], [204, Eq. (6.1.14), p. 197], and [211, Eq. (42), p. 95]. The work in [164] was motivated by F. Garvan's comments and conjectured formula for Δ^2 in [78] as a 3 by 3 Hankel determinant of classical Eisenstein series after seeing an earlier version of the present paper. Finally, the paper [165] derives a new formula for the modular discriminant Δ and a corresponding new formula that expresses Ramanujan's tau function as the difference of two positive integers. These integers arise naturally in areas as diverse as affine Lie super-algebras, higher-dimensional unimodular lattices, combinatorics, and number theory. The analysis in [165] utilizes special cases of the methods in Section 5 to extend part of the work of Jacobi in [117].

2. Fourier expansions and Lambert series

In this section we compare Fourier and Maclaurin series expansions of various ratios $f_1(u,k)$ of Jacobi elliptic functions to derive the Lambert series formulas we need. We first recall the classical Fourier expansions from [93, 117, 134, 250], write them as a double sum as in [15,

16, 21, 38, 259], and finally equate coefficients with the corresponding Maclaurin series to obtain the Lambert series formulas. At the end of this section we point out the relationship between the Lambert series $U_{2m-1}(q)$, $G_{2m+1}(q)$, $C_{2m-1}(q)$, $D_{2m+1}(q)$ appearing in our sums of squares and sums of triangles identities and the Fourier expansions of the classical Eisenstein series $E_n(\tau)$, with $q := \exp(2\pi i\tau)$ and n an even positive integer, as given by [20, p. 318] and [204, pp. 194–195].

We utilize the Jacobi elliptic function parameter

$$z := {}_2F_1\left[\begin{matrix}\frac{1}{2},\frac{1}{2}\\1\end{matrix}\,\middle|\,k^2\right] = 2K(k)/\pi \equiv 2K/\pi, \tag{2.1}$$

with

$$K(k) \equiv K := \int_0^1 \frac{dt}{\sqrt{(1-t^2)(1-k^2t^2)}} = \frac{\pi}{2}\,{}_2F_1\left[\begin{matrix}\frac{1}{2},\frac{1}{2}\\1\end{matrix}\,\middle|\,k^2\right] \tag{2.2}$$

the complete elliptic integral of the first kind in [134, Eq. (3.1.3), p. 51], and k the modulus. We also sometimes use the complete elliptic integral of the second kind

$$E(k) \equiv E := \int_0^1 \sqrt{\frac{1-k^2t^2}{1-t^2}}\,dt = \frac{\pi}{2}\,{}_2F_1\left[\begin{matrix}\frac{1}{2},-\frac{1}{2}\\1\end{matrix}\,\middle|\,k^2\right], \tag{2.3}$$

and the complementary modulus $k' := \sqrt{1-k^2}$. Finally, we take

$$q := \exp(-\pi K(\sqrt{1-k^2})/K(k)) \tag{2.4}$$

The classical Fourier expansions from [93, pp. 911–912], [117, Sections 39, 41, and 42], [134, pp. 222–225], and [250, pp. 510–520] that we need are summarized by the following theorem. Essentially all of these Fourier expansions originally appeared in [117, Sections 39, 41, and 42].

Theorem 2.1. *Let $z := 2K/\pi$, as in (2.1), with K and E given by (2.2) and (2.3), respectively. Also take $k' := \sqrt{1-k^2}$ and q as in (2.4). Then,*

$$\mathrm{sn}(u,k) = \frac{2\pi}{kK}\sum_{n=1}^{\infty}\frac{q^{n-\frac{1}{2}}}{1-q^{2n-1}}\sin\frac{(2n-1)u}{z} \tag{2.5}$$

$$\mathrm{cn}(u,k) = \frac{2\pi}{kK}\sum_{n=1}^{\infty}\frac{q^{n-\frac{1}{2}}}{1+q^{2n-1}}\cos\frac{(2n-1)u}{z} \tag{2.6}$$

$$\mathrm{dn}(u,k) = \frac{\pi}{2K} + \frac{2\pi}{K}\sum_{n=1}^{\infty}\frac{q^n}{1+q^{2n}}\cos\frac{2nu}{z} \tag{2.7}$$

$$\mathrm{sd}(u,k) = \frac{2\pi}{kk'K}\sum_{n=1}^{\infty}\frac{(-1)^{n-1}q^{n-\frac{1}{2}}}{1+q^{2n-1}}\sin\frac{(2n-1)u}{z} \tag{2.8}$$

$$\mathrm{cd}(u,k) = \frac{2\pi}{kK}\sum_{n=1}^{\infty}\frac{(-1)^{n-1}q^{n-\frac{1}{2}}}{1-q^{2n-1}}\cos\tfrac{(2n-1)u}{z} \tag{2.9}$$

$$\mathrm{nd}(u,k) = \frac{\pi}{2k'K} + \frac{2\pi}{k'K}\sum_{n=1}^{\infty}\frac{(-1)^{n}q^{n}}{1+q^{2n}}\cos\tfrac{2nu}{z} \tag{2.10}$$

$$\mathrm{sc}(u,k) = \frac{\pi}{2k'K}\tan\tfrac{u}{z} + \frac{2\pi}{k'K}\sum_{n=1}^{\infty}\frac{(-1)^{n}q^{2n}}{1+q^{2n}}\sin\tfrac{2nu}{z} \tag{2.11}$$

$$\mathrm{dc}(u,k) = \frac{\pi}{2K}\sec\tfrac{u}{z} + \frac{2\pi}{K}\sum_{n=1}^{\infty}\frac{(-1)^{n-1}q^{2n-1}}{1-q^{2n-1}}\cos\tfrac{(2n-1)u}{z} \tag{2.12}$$

$$\mathrm{nc}(u,k) = \frac{\pi}{2k'K}\sec\tfrac{u}{z} - \frac{2\pi}{k'K}\sum_{n=1}^{\infty}\frac{(-1)^{n-1}q^{2n-1}}{1+q^{2n-1}}\cos\tfrac{(2n-1)u}{z} \tag{2.13}$$

$$\mathrm{sn}^2(u,k) = \frac{K-E}{k^2K} - \frac{2\pi^2}{k^2K^2}\sum_{n=1}^{\infty}\frac{nq^{n}}{1-q^{2n}}\cos\tfrac{2nu}{z} \tag{2.14}$$

$$\mathrm{sc}^2(u,k) = -\frac{1}{k'^2}\frac{E}{K} + \frac{1}{z^2k'^2}\sec^2\tfrac{u}{z} - \frac{8}{z^2k'^2}\sum_{n=1}^{\infty}\frac{(-1)^{n}nq^{2n}}{1-q^{2n}}\cos\tfrac{2nu}{z} \tag{2.15}$$

$$\mathrm{sd}^2(u,k) = \frac{E-k'^2K}{k^2k'^2K} + \frac{8}{z^2k^2k'^2}\sum_{n=1}^{\infty}\frac{(-1)^{n}nq^{n}}{1-q^{2n}}\cos\tfrac{2nu}{z} \tag{2.16}$$

$$\frac{\mathrm{sn}(u,k)\,\mathrm{cn}(u,k)}{\mathrm{dn}(u,k)} = \frac{4\pi}{k^2K}\sum_{n=1}^{\infty}\frac{q^{2n-1}}{1-q^{4n-2}}\sin\tfrac{(4n-2)u}{z} \tag{2.17}$$

$$\frac{\mathrm{sn}(u,k)\,\mathrm{dn}(u,k)}{\mathrm{cn}(u,k)} = \frac{\pi}{2K}\tan\tfrac{u}{z} + \frac{2\pi}{K}\sum_{n=1}^{\infty}\frac{q^{n}}{1+(-1)^{n}q^{n}}\sin\tfrac{2nu}{z} \tag{2.18}$$

$$\frac{\mathrm{sn}(u,k)}{\mathrm{cn}(u,k)\,\mathrm{dn}(u,k)} = \frac{\pi}{2k'^2K}\tan\tfrac{u}{z} + \frac{2\pi}{k'^2K}\sum_{n=1}^{\infty}\frac{(-1)^{n}q^{n}}{1+q^{n}}\sin\tfrac{2nu}{z} \tag{2.19}$$

$$\frac{\mathrm{sn}^2(u,k)\,\mathrm{cn}^2(u,k)}{\mathrm{dn}^2(u,k)} = \frac{1}{k^4} + \frac{k'^2}{k^4} - \frac{2}{k^4}\frac{E}{K} - \frac{32}{z^2k^4}\sum_{n=1}^{\infty}\frac{nq^{2n}}{1-q^{4n}}\cos\tfrac{4nu}{z} \tag{2.20}$$

$$\frac{\mathrm{sn}^2(u,k)\,\mathrm{dn}^2(u,k)}{\mathrm{cn}^2(u,k)} = 1 - 2\frac{E}{K} + \frac{1}{z^2}\sec^2\tfrac{u}{z} - \frac{8}{z^2}\sum_{n=1}^{\infty}\frac{nq^{n}}{1-(-1)^{n}q^{n}}\cos\tfrac{2nu}{z} \tag{2.21}$$

$$\frac{\mathrm{sn}^2(u,k)}{\mathrm{cn}^2(u,k)\,\mathrm{dn}^2(u,k)} = \frac{1}{k'^2} - \frac{2}{k'^4}\frac{E}{K} + \frac{1}{z^2k'^4}\sec^2\tfrac{u}{z}$$

$$- \frac{8}{z^2k'^4}\sum_{n=1}^{\infty}\frac{(-1)^{n}nq^{n}}{1-q^{n}}\cos\tfrac{2nu}{z} \tag{2.22}$$

$$\mathrm{sn}(u,k)\,\mathrm{dn}(u,k) = \frac{\pi^2}{kK^2}\sum_{n=1}^{\infty}\frac{(2n-1)q^{n-\frac{1}{2}}}{1+q^{2n-1}}\sin\tfrac{(2n-1)u}{z} \tag{2.23}$$

$$\operatorname{sn}(u,k)\operatorname{cn}(u,k) = \frac{2\pi^2}{k^2K^2}\sum_{n=1}^{\infty}\frac{nq^n}{1+q^{2n}}\sin\tfrac{2nu}{z} \tag{2.24}$$

$$\frac{\operatorname{sn}(u,k)}{\operatorname{dn}^2(u,k)} = \frac{\pi^2}{kk'^2K^2}\sum_{n=1}^{\infty}\frac{(-1)^{n-1}(2n-1)q^{n-\frac{1}{2}}}{1-q^{2n-1}}\sin\tfrac{(2n-1)u}{z} \tag{2.25}$$

$$\frac{\operatorname{sn}(u,k)\operatorname{cn}(u,k)}{\operatorname{dn}^2(u,k)} = -\frac{2\pi^2}{k^2k'K^2}\sum_{n=1}^{\infty}\frac{(-1)^n nq^n}{1+q^{2n}}\sin\tfrac{2nu}{z} \tag{2.26}$$

$$\frac{\operatorname{sn}(u,k)}{\operatorname{cn}^2(u,k)} = \frac{1}{z^2k'^2}\sec\tfrac{u}{z}\tan\tfrac{u}{z} - \frac{\pi^2}{K^2}\sum_{n=1}^{\infty}\frac{(-1)^{n-1}(2n-1)q^{2n-1}}{1-q^{2n-1}} \times \sin\tfrac{(2n-1)u}{z} \tag{2.27}$$

$$\frac{\operatorname{sn}(u,k)\operatorname{dn}(u,k)}{\operatorname{cn}^2(u,k)} = \frac{\pi^2}{4k'K^2}\sec\tfrac{u}{z}\tan\tfrac{u}{z} + \frac{\pi^2}{k'K^2}\sum_{n=1}^{\infty}\frac{(-1)^{n-1}(2n-1)q^{2n-1}}{1+q^{2n-1}}\sin\tfrac{(2n-1)u}{z} \tag{2.28}$$

The Fourier series expansions in (2.18) and (2.21) are important ingredients of our derivation of the $4n^2$ and $4n(n+1)$ squares identities, respectively, in Section 5. Both of these expansions are direct consequences of simpler Fourier expansions.

A derivation of (2.18) is outlined in [134, Example 6, p. 243]. The Fourier expansions for $\operatorname{ns}(2u,k)$ and $\operatorname{cs}(2u,k)$ are substituted into

$$\operatorname{sn}(u,k)\operatorname{dc}(u,k) = \operatorname{ns}(2u,k) - \operatorname{cs}(2u,k). \tag{2.29}$$

The terms from the sum in (2.18) with n odd come from $\operatorname{ns}(2u,k)$, and with n even from $\operatorname{cs}(2u,k)$. The $\tan\frac{u}{z}$ in (2.18) follows from $\csc\theta - \cot\theta = \tan\frac{\theta}{2}$.

In order to derive (2.21) we first apply the Pythagorean relations for $\operatorname{dn}^2(u,k)$ and $\operatorname{sn}^2(u,k)$ to obtain

$$\begin{aligned}\operatorname{sc}^2(u,k)\operatorname{dn}^2(u,k) &= \operatorname{sc}^2(u,k)[1-k^2\operatorname{sn}^2(u,k)]\\ &= \operatorname{sc}^2(u,k)[1-k^2(1-\operatorname{cn}^2(u,k))]\\ &= (1-k^2)\operatorname{sc}^2(u,k)+k^2\operatorname{sn}^2(u,k)\end{aligned} \tag{2.30}$$

Next, substitute (2.14) and (2.15) into (2.30). Finally, combine the two resulting sums termwise while appealing to the difference of squares

$$(1-q^{2n}) = (1+(-1)^n q^n)(1-(-1)^n q^n). \tag{2.31}$$

This last step is motivated by [94, p. 119].

The Fourier series expansions in (2.23)–(2.28) are obtained by differentiating those in (2.6), (2.7), (2.9), (2.10), (2.12), and (2.13).

The Fourier expansions in Theorem 2.1 may be written as a double sum by first expanding the $\sin\frac{Nu}{z}$ and $\cos\frac{Nu}{z}$ as Maclaurin series, interchanging summation, and then simplifying. We obtain

Theorem 2.2. *Let $z := 2K/\pi$, as in (2.1), with K and E given by (2.2) and (2.3), respectively. Also take $k' := \sqrt{1-k^2}$ and q as in (2.4). Then,*

$$zk \cdot \mathrm{sn}(u,k) = 4\sum_{m=1}^{\infty} \frac{(-1)^{m-1}}{z^{2m-1}} \left[\sum_{r=1}^{\infty} \frac{(2r-1)^{2m-1} q^{r-\frac{1}{2}}}{1-q^{2r-1}}\right] \frac{u^{2m-1}}{(2m-1)!} \tag{2.32}$$

$$zk \cdot \mathrm{cn}(u,k) = 4\sum_{m=0}^{\infty} \frac{(-1)^{m}}{z^{2m}} \left[\sum_{r=1}^{\infty} \frac{(2r-1)^{2m} q^{r-\frac{1}{2}}}{1+q^{2r-1}}\right] \frac{u^{2m}}{(2m)!} \tag{2.33}$$

$$z \cdot \mathrm{dn}(u,k) = 1 + 4\sum_{m=0}^{\infty} \frac{(-1)^{m}2^{2m}}{z^{2m}} \left[\sum_{r=1}^{\infty} \frac{r^{2m} q^{r}}{1+q^{2r}}\right] \frac{u^{2m}}{(2m)!} \tag{2.34}$$

$$zkk' \cdot \mathrm{sd}(u,k) = 4\sum_{m=1}^{\infty} \frac{(-1)^{m-1}}{z^{2m-1}} \left[\sum_{r=1}^{\infty} \frac{(-1)^{r+1}(2r-1)^{2m-1} q^{r-\frac{1}{2}}}{1+q^{2r-1}}\right] \frac{u^{2m-1}}{(2m-1)!} \tag{2.35}$$

$$zk \cdot \mathrm{cd}(u,k) = 4\sum_{m=0}^{\infty} \frac{(-1)^{m}}{z^{2m}} \left[\sum_{r=1}^{\infty} \frac{(-1)^{r+1}(2r-1)^{2m} q^{r-\frac{1}{2}}}{1-q^{2r-1}}\right] \frac{u^{2m}}{(2m)!} \tag{2.36}$$

$$zk' \cdot \mathrm{nd}(u,k) = 1 + 4\sum_{m=0}^{\infty} \frac{(-1)^{m}2^{2m}}{z^{2m}} \left[\sum_{r=1}^{\infty} \frac{(-1)^{r} r^{2m} q^{r}}{1+q^{2r}}\right] \frac{u^{2m}}{(2m)!} \tag{2.37}$$

$$zk' \cdot \mathrm{sc}(u,k) = \tan\tfrac{u}{z} + 4\sum_{m=1}^{\infty} \frac{(-1)^{m-1}2^{2m-1}}{z^{2m-1}} \times \left[\sum_{r=1}^{\infty} \frac{(-1)^{r} r^{2m-1} q^{2r}}{1+q^{2r}}\right] \frac{u^{2m-1}}{(2m-1)!} \tag{2.38}$$

$$z \cdot \mathrm{dc}(u,k) = \sec\tfrac{u}{z} + 4\sum_{m=0}^{\infty} \frac{(-1)^{m}}{z^{2m}} \times \left[\sum_{r=1}^{\infty} \frac{(-1)^{r+1}(2r-1)^{2m} q^{2r-1}}{1-q^{2r-1}}\right] \frac{u^{2m}}{(2m)!} \tag{2.39}$$

$$zk' \cdot \mathrm{nc}(u,k) = \sec\tfrac{u}{z} - 4\sum_{m=0}^{\infty} \frac{(-1)^{m}}{z^{2m}} \times \left[\sum_{r=1}^{\infty} \frac{(-1)^{r-1}(2r-1)^{2m} q^{2r-1}}{1+q^{2r-1}}\right] \frac{u^{2m}}{(2m)!} \tag{2.40}$$

$$z^2k^2 \cdot \mathrm{sn}^2(u,k) = z^2 - z^2\frac{E}{K} - 8\sum_{m=0}^{\infty} \frac{(-1)^{m}2^{2m}}{z^{2m}} \left[\sum_{r=1}^{\infty} \frac{r^{2m+1} q^{r}}{1-q^{2r}}\right] \frac{u^{2m}}{(2m)!} \tag{2.41}$$

$$z^2k'^2 \cdot \mathrm{sc}^2(u,k) = -z^2\frac{E}{K} + \sec^2\tfrac{u}{z} + 8\sum_{m=0}^{\infty}\frac{(-1)^m 2^{2m}}{z^{2m}} \times \left[\sum_{r=1}^{\infty}\frac{(-1)^{r-1}r^{2m+1}q^{2r}}{1-q^{2r}}\right]\frac{u^{2m}}{(2m)!} \tag{2.42}$$

$$z^2k^2k'^2 \cdot \mathrm{sd}^2(u,k) = -z^2k'^2 + z^2\frac{E}{K} - 8\sum_{m=0}^{\infty}\frac{(-1)^m 2^{2m}}{z^{2m}} \times \left[\sum_{r=1}^{\infty}\frac{(-1)^{r+1}r^{2m+1}q^{r}}{1-q^{2r}}\right]\frac{u^{2m}}{(2m)!} \tag{2.43}$$

$$zk^2 \cdot \frac{\mathrm{sn}(u,k)\,\mathrm{cn}(u,k)}{\mathrm{dn}(u,k)} = 8\sum_{m=1}^{\infty}\frac{(-1)^{m-1}2^{2m-1}}{z^{2m-1}}\left[\sum_{r=1}^{\infty}\frac{(2r-1)^{2m-1}q^{2r-1}}{1-q^{4r-2}}\right]\frac{u^{2m-1}}{(2m-1)!} \tag{2.44}$$

$$z \cdot \frac{\mathrm{sn}(u,k)\,\mathrm{dn}(u,k)}{\mathrm{cn}(u,k)} = \tan\tfrac{u}{z} + 4\sum_{m=1}^{\infty}\frac{(-1)^{m-1}2^{2m-1}}{z^{2m-1}} \times \left[\sum_{r=1}^{\infty}\frac{r^{2m-1}q^{r}}{1+(-1)^r q^{r}}\right]\frac{u^{2m-1}}{(2m-1)!} \tag{2.45}$$

$$zk'^2 \cdot \frac{\mathrm{sn}(u,k)}{\mathrm{cn}(u,k)\,\mathrm{dn}(u,k)} = \tan\tfrac{u}{z} + 4\sum_{m=1}^{\infty}\frac{(-1)^{m-1}2^{2m-1}}{z^{2m-1}} \times \left[\sum_{r=1}^{\infty}\frac{(-1)^r r^{2m-1}q^{r}}{1+q^{r}}\right]\frac{u^{2m-1}}{(2m-1)!} \tag{2.46}$$

$$z^2k^4 \cdot \frac{\mathrm{sn}^2(u,k)\,\mathrm{cn}^2(u,k)}{\mathrm{dn}^2(u,k)} = z^2 + z^2k'^2 - 2z^2\frac{E}{K} - 32\sum_{m=0}^{\infty}\frac{(-1)^m 2^{4m}}{z^{2m}}\left[\sum_{r=1}^{\infty}\frac{r^{2m+1}q^{2r}}{1-q^{4r}}\right]\frac{u^{2m}}{(2m)!} \tag{2.47}$$

$$z^2 \cdot \frac{\mathrm{sn}^2(u,k)\,\mathrm{dn}^2(u,k)}{\mathrm{cn}^2(u,k)} = z^2 - 2z^2\frac{E}{K} + \sec^2\tfrac{u}{z} - 8\sum_{m=0}^{\infty}\frac{(-1)^m 2^{2m}}{z^{2m}}\left[\sum_{r=1}^{\infty}\frac{r^{2m+1}q^{r}}{1-(-1)^r q^{r}}\right]\frac{u^{2m}}{(2m)!} \tag{2.48}$$

$$z^2k'^4 \cdot \frac{\mathrm{sn}^2(u,k)}{\mathrm{cn}^2(u,k)\,\mathrm{dn}^2(u,k)} = z^2k'^2 - 2z^2\frac{E}{K} + \sec^2\tfrac{u}{z} + 8\sum_{m=0}^{\infty}\frac{(-1)^m 2^{2m}}{z^{2m}} \times \left[\sum_{r=1}^{\infty}\frac{(-1)^{r-1}r^{2m+1}q^{r}}{1-q^{r}}\right]\frac{u^{2m}}{(2m)!} \tag{2.49}$$

$$z^2k\cdot\mathrm{sn}(u,k)\,\mathrm{dn}(u,k)=4\sum_{m=1}^{\infty}\frac{(-1)^{m-1}}{z^{2m-1}}\left[\sum_{r=1}^{\infty}\frac{(2r-1)^{2m}q^{r-\frac{1}{2}}}{1+q^{2r-1}}\right]\frac{u^{2m-1}}{(2m-1)!} \tag{2.50}$$

$$z^2k^2\cdot\mathrm{sn}(u,k)\,\mathrm{cn}(u,k)=8\sum_{m=1}^{\infty}\frac{(-1)^{m-1}2^{2m-1}}{z^{2m-1}}\left[\sum_{r=1}^{\infty}\frac{r^{2m}q^{r}}{1+q^{2r}}\right]\frac{u^{2m-1}}{(2m-1)!} \tag{2.51}$$

$$z^2kk'^2\cdot\frac{\mathrm{sn}(u,k)}{\mathrm{dn}^2(u,k)}=4\sum_{m=1}^{\infty}\frac{(-1)^{m-1}}{z^{2m-1}}\left[\sum_{r=1}^{\infty}\frac{(-1)^{r+1}(2r-1)^{2m}q^{r-\frac{1}{2}}}{1-q^{2r-1}}\right]\frac{u^{2m-1}}{(2m-1)!} \tag{2.52}$$

$$z^2k'k^2\cdot\frac{\mathrm{sn}(u,k)\,\mathrm{cn}(u,k)}{\mathrm{dn}^2(u,k)}=-8\sum_{m=1}^{\infty}\frac{(-1)^{m-1}2^{2m-1}}{z^{2m-1}}\left[\sum_{r=1}^{\infty}\frac{(-1)^{r}r^{2m}q^{r}}{1+q^{2r}}\right]\frac{u^{2m-1}}{(2m-1)!} \tag{2.53}$$

$$z^2k'^2\cdot\frac{\mathrm{sn}(u,k)}{\mathrm{cn}^2(u,k)}=\sec\tfrac{u}{z}\tan\tfrac{u}{z}-4\sum_{m=1}^{\infty}\frac{(-1)^{m-1}}{z^{2m-1}}\times\left[\sum_{r=1}^{\infty}\frac{(-1)^{r+1}(2r-1)^{2m}q^{2r-1}}{1-q^{2r-1}}\right]\frac{u^{2m-1}}{(2m-1)!} \tag{2.54}$$

$$z^2k'\cdot\frac{\mathrm{sn}(u,k)\,\mathrm{dn}(u,k)}{\mathrm{cn}^2(u,k)}=\sec\tfrac{u}{z}\tan\tfrac{u}{z}+4\sum_{m=1}^{\infty}\frac{(-1)^{m-1}}{z^{2m-1}}\times\left[\sum_{r=1}^{\infty}\frac{(-1)^{r+1}(2r-1)^{2m}q^{2r-1}}{1+q^{2r-1}}\right]\frac{u^{2m-1}}{(2m-1)!} \tag{2.55}$$

In order to equate coefficients of u^N in Theorem 2.2 we first need the expansions [93, p. 35] of $\tan\frac{u}{z}$, $\sec^2\frac{u}{z}$, $\sec\frac{u}{z}$, and $\sec\frac{u}{z}\tan\frac{u}{z}$ given by

$$\tan\tfrac{u}{z}=\sum_{m=1}^{\infty}\frac{2^{2m}(2^{2m}-1)|B_{2m}|}{(2m)!\,z^{2m-1}}u^{2m-1},\quad\text{for }\frac{u^2}{z^2}<\frac{\pi^2}{4}, \tag{2.56}$$

$$\sec^2\tfrac{u}{z}=\sum_{m=0}^{\infty}\frac{2^{2m+1}(2^{2m+2}-1)|B_{2m+2}|}{(m+1)(2m)!\,z^{2m}}u^{2m}, \tag{2.57}$$

$$\sec\tfrac{u}{z}=\sum_{m=0}^{\infty}\frac{|E_{2m}|}{(2m)!\,z^{2m}}u^{2m},\quad\text{for }\frac{u^2}{z^2}<\frac{\pi^2}{4}, \tag{2.58}$$

$$\sec\tfrac{u}{z}\tan\tfrac{u}{z}=\sum_{m=1}^{\infty}\frac{|E_{2m}|}{(2m-1)!\,z^{2m-1}}u^{2m-1}, \tag{2.59}$$

where the Bernoulli numbers B_n and Euler numbers E_n are defined in [48, pp. 48–49] by

$$\frac{t}{e^t-1}:=\sum_{n=0}^{\infty}B_n\frac{t^n}{n!},\quad\text{for }|t|<2\pi, \tag{2.60}$$

and

$$\frac{2e^t}{e^{2t}+1} := \sum_{n=0}^{\infty} E_n \frac{t^n}{n!}, \quad \text{for } |t| < \pi/2. \tag{2.61}$$

Convenient, explicit formulas for the Bernoulli numbers B_n can be found in [91].

We next write down the Maclaurin series expansions of the ratios of Jacobi elliptic functions in Theorem 2.2. The coefficients of u^N in these expansions are polynomials in k^2, where k is the modulus. We have

Definition 2.3 (Maclaurin series expansion polynomials for Jacobi elliptic functions). Let the elliptic function polynomials $(\text{elliptic})_m(k^2)$ of k^2, with k the modulus, be determined by the coefficients of the following Maclaurin series expansions:

$$\text{sn}(u,k) = \sum_{m=1}^{\infty} (sn)_m(k^2)\frac{u^{2m-1}}{(2m-1)!}; \quad \text{cn}(u,k) = \sum_{m=0}^{\infty} (cn)_m(k^2)\frac{u^{2m}}{(2m)!}, \tag{2.62}$$

$$\text{dn}(u,k) = \sum_{m=0}^{\infty} (dn)_m(k^2)\frac{u^{2m}}{(2m)!}; \quad \text{sd}(u,k) = \sum_{m=1}^{\infty} (s/d)_m(k^2)\frac{u^{2m-1}}{(2m-1)!}, \tag{2.63}$$

$$\text{cd}(u,k) = \sum_{m=0}^{\infty} (c/d)_m(k^2)\frac{u^{2m}}{(2m)!}; \quad \text{nd}(u,k) = \sum_{m=0}^{\infty} (nd)_m(k^2)\frac{u^{2m}}{(2m)!}, \tag{2.64}$$

$$\text{sc}(u,k) = \sum_{m=1}^{\infty} (s/c)_m(k^2)\frac{u^{2m-1}}{(2m-1)!}; \quad \text{dc}(u,k) = \sum_{m=0}^{\infty} (d/c)_m(k^2)\frac{u^{2m}}{(2m)!}, \tag{2.65}$$

$$\text{nc}(u,k) = \sum_{m=0}^{\infty} (nc)_m(k^2)\frac{u^{2m}}{(2m)!}; \quad \text{sn}^2(u,k) = \sum_{m=1}^{\infty} (sn^2)_m(k^2)\frac{u^{2m}}{(2m)!}, \tag{2.66}$$

$$\text{sc}^2(u,k) = \sum_{m=1}^{\infty} (s^2/c^2)_m(k^2)\frac{u^{2m}}{(2m)!}; \quad \text{sd}^2(u,k) = \sum_{m=1}^{\infty} (s^2/d^2)_m(k^2)\frac{u^{2m}}{(2m)!}, \tag{2.67}$$

$$\frac{\text{sn}(u,k)\,\text{cn}(u,k)}{\text{dn}(u,k)} = \sum_{m=1}^{\infty} (sc/d)_m(k^2)\frac{u^{2m-1}}{(2m-1)!}, \tag{2.68}$$

$$\frac{\text{sn}(u,k)\,\text{dn}(u,k)}{\text{cn}(u,k)} = \sum_{m=1}^{\infty} (sd/c)_m(k^2)\frac{u^{2m-1}}{(2m-1)!}, \tag{2.69}$$

$$\frac{\text{sn}(u,k)}{\text{cn}(u,k)\,\text{dn}(u,k)} = \sum_{m=1}^{\infty} (s/cd)_m(k^2)\frac{u^{2m-1}}{(2m-1)!}, \tag{2.70}$$

$$\frac{\text{sn}^2(u,k)\,\text{cn}^2(u,k)}{\text{dn}^2(u,k)} = \sum_{m=1}^{\infty} (s^2c^2/d^2)_m(k^2)\frac{u^{2m}}{(2m)!}, \tag{2.71}$$

$$\frac{\text{sn}^2(u,k)\,\text{dn}^2(u,k)}{\text{cn}^2(u,k)} = \sum_{m=1}^{\infty} (s^2d^2/c^2)_m(k^2)\frac{u^{2m}}{(2m)!}, \tag{2.72}$$

$$\frac{\operatorname{sn}^2(u,k)}{\operatorname{cn}^2(u,k)\operatorname{dn}^2(u,k)} = \sum_{m=1}^{\infty}(s^2/c^2d^2)_m(k^2)\frac{u^{2m}}{(2m)!}, \tag{2.73}$$

$$\operatorname{sn}(u,k)\operatorname{dn}(u,k) = \sum_{m=1}^{\infty}(sd)_m(k^2)\frac{u^{2m-1}}{(2m-1)!}, \tag{2.74}$$

$$\operatorname{sn}(u,k)\operatorname{cn}(u,k) = \sum_{m=1}^{\infty}(sc)_m(k^2)\frac{u^{2m-1}}{(2m-1)!}, \tag{2.75}$$

$$\frac{\operatorname{sn}(u,k)}{\operatorname{dn}^2(u,k)} = \sum_{m=1}^{\infty}(s/d^2)_m(k^2)\frac{u^{2m-1}}{(2m-1)!}, \tag{2.76}$$

$$\frac{\operatorname{sn}(u,k)\operatorname{cn}(u,k)}{\operatorname{dn}^2(u,k)} = \sum_{m=1}^{\infty}(sc/d^2)_m(k^2)\frac{u^{2m-1}}{(2m-1)!}, \tag{2.77}$$

$$\frac{\operatorname{sn}(u,k)}{\operatorname{cn}^2(u,k)} = \sum_{m=1}^{\infty}(s/c^2)_m(k^2)\frac{u^{2m-1}}{(2m-1)!}, \tag{2.78}$$

$$\frac{\operatorname{sn}(u,k)\operatorname{dn}(u,k)}{\operatorname{cn}^2(u,k)} = \sum_{m=1}^{\infty}(sd/c^2)_m(k^2)\frac{u^{2m-1}}{(2m-1)!}. \tag{2.79}$$

Recursions, explicit computations, and tables of the polynomials $(\text{elliptic})_m(k^2)$ in the simpler expansions above such as for sn, cn, dn, and sn^2 in (2.62), (2.63), and (2.66) can be found in [15–18, 21, 31, 49, 50, 58–60, 93, 117, 134, 170, 186, 188, 208, 209, 250, 252, 253, 259]. Corresponding applications to combinatorics appear in [21, 54, 58–60, 69–71, 208, 209, 224, 244, 252, 253, 257].

Keeping in mind (2.56)–(2.61) and Definition 2.3, we find that equating coefficients of u^N in Theorem 2.2 yields the Lambert series formulas in

Theorem 2.4. *Let $z := 2K(k)/\pi \equiv 2K/\pi$, as in (2.1), with k the modulus. Let q be as in (2.4), and let the Bernoulli numbers B_n and Euler numbers E_n be defined by (2.60) and (2.61), respectively. Take $(sd/c)_m(k^2)$, $(s^2d^2/c^2)_m(k^2)$, $(nc)_{m-1}(k^2)$, $(sc/d)_m(k^2)$, $(sn^2)_m(k^2)$, $(cn)_{m-1}(k^2)$, $(dn)_m(k^2)$, $(sd)_m(k^2)$, $(sc)_m(k^2)$, and $(sd/c^2)_m(k^2)$ to be the elliptic function polynomials of k^2 determined by Definition 2.3. Let $m = 1, 2, 3, \ldots$. Then,*

$$\begin{aligned} U_{2m-1}(-q) &:= \sum_{r=1}^{\infty}\frac{-r^{2m-1}q^r}{1+(-1)^r q^r} \\ &= (-1)^{m-1}\frac{(2^{2m}-1)}{4m}\cdot|B_{2m}| + (-1)^m\frac{z^{2m}}{2^{2m+1}}\cdot(sd/c)_m(k^2), \end{aligned} \tag{2.80}$$

$$\begin{aligned} G_{2m+1}(-q) &:= \sum_{r=1}^{\infty}\frac{r^{2m+1}q^r}{1-(-1)^r q^r} \\ &= (-1)^m\frac{(2^{2m+2}-1)}{4(m+1)}\cdot|B_{2m+2}| + (-1)^{m-1}\frac{z^{2m+2}}{2^{2m+3}}\cdot(s^2d^2/c^2)_m(k^2), \end{aligned} \tag{2.81}$$

$$R_{2m-2}(q) := \sum_{r=1}^{\infty}(-1)^{r+1}\frac{(2r-1)^{2m-2}q^{2r-1}}{1+q^{2r-1}}$$
$$= (-1)^{m-1}\cdot\tfrac{1}{4}\cdot|E_{2m-2}| + (-1)^m\frac{z^{2m-1}}{4}\sqrt{1-k^2}\cdot(nc)_{m-1}(k^2), \tag{2.82}$$

$$C_{2m-1}(q) := \sum_{r=1}^{\infty}\frac{(2r-1)^{2m-1}q^{2r-1}}{1-q^{2(2r-1)}} = (-1)^{m-1}\frac{z^{2m}k^2}{2^{2m+2}}\cdot(sc/d)_m(k^2), \tag{2.83}$$

$$D_{2m+1}(q) := \sum_{r=1}^{\infty}\frac{r^{2m+1}q^r}{1-q^{2r}} = (-1)^{m-1}\frac{z^{2m+2}k^2}{2^{2m+3}}\cdot(sn^2)_m(k^2), \tag{2.84}$$

$$T_{2m-2}(q) := \sum_{r=1}^{\infty}\frac{(2r-1)^{2m-2}q^{r-\frac{1}{2}}}{1+q^{2r-1}} = (-1)^{m-1}\frac{z^{2m-1}k}{4}\cdot(cn)_{m-1}(k^2), \tag{2.85}$$

$$N_{2m}(q) := \sum_{r=1}^{\infty}\frac{r^{2m}q^r}{1+q^{2r}} = (-1)^m\frac{z^{2m+1}}{2^{2m+2}}\cdot(dn)_m(k^2), \tag{2.86}$$

and

$$N_0(q) := \sum_{r=1}^{\infty}\frac{q^r}{1+q^{2r}} = -\tfrac{1}{4}+\frac{z}{4}\cdot(dn)_0(k^2) = -\tfrac{1}{4}+\frac{z}{4}, \tag{2.87}$$

$$T_{2m}(q) = (-1)^{m+1}\frac{z^{2m+1}k}{4}\cdot(sd)_m(k^2), \tag{2.88}$$

$$N_{2m}(q) = (-1)^{m+1}\frac{z^{2m+1}k^2}{2^{2m+2}}\cdot(sc)_m(k^2), \tag{2.89}$$

$$R_{2m}(q) = (-1)^m\cdot\tfrac{1}{4}\cdot|E_{2m}| + (-1)^{m+1}\frac{z^{2m+1}}{4}\sqrt{1-k^2}\cdot(sd/c^2)_m(k^2). \tag{2.90}$$

The Lambert series in (2.32) and (2.35) are constant multiples of the one in (2.44), with $q \mapsto q^{1/2}$ and $q \mapsto i\sqrt{q}$, respectively. The Lambert series in (2.33), (2.34), (2.40), (2.41), (2.45), (2.48) are transformed by $q \mapsto -q$ into the corresponding Lambert series in (2.36), (2.37), (2.39), (2.43), (2.46), (2.49), respectively. The substitution $q \mapsto -q^2$ transforms the Lambert series in (2.45) and (2.48) into those in (2.38) and (2.42), respectively. Taking $q \mapsto q^2$ transforms the Lambert series in (2.41) into the one in (2.47). The relations in (2.88), (2.89), and (2.90) also follow from combining (2.85), (2.86), and (2.82) with the Maclaurin series expansions of the derivative formulas for $\mathrm{cn}(u,k)$, $\mathrm{dn}(u,k)$, and $\mathrm{nc}(u,k)$. The above substitutions, combined with the corresponding modular transformations, is why we only need the 10 Lambert series formulas in Theorem 2.4.

The analysis leading to Theorem 2.4 is similar to that in [15, 16, 21, 23, 38, 259]. The Lambert series identities in (2.80)–(2.86) are equivalent to the identities in Tables 1(x), 1(ii), 1(xiv), 1(vii), 1(iii), 1(xv), 1(xi), respectively, of [259]. In this paper we mainly work with the above rational functions of $\mathrm{sn}(u,k)$, $\mathrm{cn}(u,k)$, and $\mathrm{dn}(u,k)$ which *do not* have $\mathrm{sn}(u,k)$ as a factor of the denominator. Our analysis in Sections 3 and 4 of classical continued fraction expansions of the Laplace transforms of these rational functions leads to our elegant product formulas for Hankel and χ determinants that are crucial in Section 5.

Table 1 of Zucker's paper [259] also contains formulas for the Lambert series determined by the Fourier series expansions of rational functions, analogous to several of those in Definition 2.3, of $\operatorname{sn}(u,k)$, $\operatorname{cn}(u,k)$, and $\operatorname{dn}(u,k)$, that *do* have $\operatorname{sn}(u,k)$ as a factor of the denominator. This second class of rational functions (and their corresponding Lambert series) does not lead to simple, "closed" product formulas for Hankel and χ determinants.

The Lambert series $V_s(q)$ in (1.18) is one such example. Moreover, a formula for $V_s(q^2)$, corresponding to $\operatorname{ns}^2(u,k)$, appears in Table 1(i) of [259]. Even so, in [164], we discuss how the Fourier series expansion of $\operatorname{ns}^2(u,k)$ has applications to Hankel determinants of classical Eisenstein series.

We conclude this section with a discussion of the relationship between the Lambert series $U_{2m-1}(q), G_{2m+1}(q), C_{2m-1}(q), D_{2m+1}(q)$ and the Fourier expansions of the classical Eisenstein series $E_n(\tau)$ as given by [20, p. 318] and [204, pp. 194–195] in the following definition.

Definition 2.5. Let $q := \exp(2\pi i\tau)$, where τ is in the upper half-plane $\mathcal{H}$, and take $y := \operatorname{Im}(\tau) > 0$. Let $n = 1, 2, 3, \ldots$. We then have

$$E_2(\tau) \equiv E_2(q) := 1 - 24\sum_{r=1}^{\infty}\frac{rq^r}{1-q^r} - \frac{3}{\pi y}, \tag{2.91}$$

and for $n \geq 2$,

$$E_{2n}(\tau) \equiv E_{2n}(q) := 1 - \frac{4n}{B_{2n}}\sum_{r=1}^{\infty}\frac{r^{2n-1}q^r}{1-q^r}, \tag{2.92}$$

with the B_{2n} the Bernoulli numbers in (2.60).

As examples, Ramanujan in [194] studied the series $L := E_2(\tau) + 3/\pi y$, $M := E_4(\tau)$, and $N := E_6(\tau)$. Note that we equate the q's in (2.4) and Definition 2.5. That is $2\tau = i\boldsymbol{K}(\sqrt{1-k^2})/\boldsymbol{K}(k)$.

We now write our Lambert series $U_{2m-1}(q), G_{2m+1}(q), C_{2m-1}(q)$, and $D_{2m+1}(q)$ as linear combinations of the $E_n(\tau)$ in the following lemma.

Lemma 2.6. *Let $U_{2m-1}(q), G_{2m+1}(q), C_{2m-1}(q)$, and $D_{2m+1}(q)$ be determined by* (2.80), (2.81), (2.83), *and* (2.84), *respectively, with q as in* (2.4). *Take $E_{2n}(q)$ as in Definition* 2.5, *with $2\tau = i\boldsymbol{K}(\sqrt{1-k^2})/\boldsymbol{K}(k)$. Let $m = 1, 2, 3, \ldots$. We then have*

$$U_{2m-1}(q) = \frac{B_{2m}}{4m}\{(2^{2m}-1) - E_{2m}(q) + 2(1+2^{2m-1})E_{2m}(q^2) - 2^{2m+1}E_{2m}(q^4)\}, \tag{2.93}$$

$$G_{2m+1}(q) = \frac{B_{2m+2}}{4(m+1)}\{(2^{2m+2}-1) + E_{2m+2}(q) - 2^{2m+2}E_{2m+2}(q^2)\}, \tag{2.94}$$

$$C_{2m-1}(q) = \frac{B_{2m}}{4m}\{-E_{2m}(q) + (1+2^{2m-1})E_{2m}(q^2) - 2^{2m-1}E_{2m}(q^4)\}, \tag{2.95}$$

$$D_{2m+1}(q) = \frac{B_{2m+2}}{4(m+1)}\{-E_{2m+2}(q) + E_{2m+2}(q^2)\}, \tag{2.96}$$

with B_{2n} the Bernoulli numbers defined by (2.60).

Proof: To establish (2.93), consider the elementary identity

$$\sum_{r=1}^{\infty}(-1)^r f(r) = 2\sum_{r=1}^{\infty} f(2r) - \sum_{r=1}^{\infty} f(r), \tag{2.97}$$

with $f(r)$ determined by the $q \mapsto -q$ case of (2.80). Apply the trivial identity

$$\frac{x}{1+x} = \frac{x}{1-x} - \frac{2x^2}{1-x^2} \tag{2.98}$$

termwise to the two resulting sums. Finally, use (2.91) or (2.92) to solve for each of the next four sums in terms of the E_{2m}, and simplify. For the case $m = 1$, note that y, $2y$, and $4y$ correspond to q, q^2, and q^4, respectively, and that $(-3/\pi y) + (18/2\pi y) + (-24/4\pi y) = 0$.

Equation (2.94) is an immediate consequence of (2.97), with $f(r)$ determined by the $q \mapsto -q$ case of (2.81), using (2.92) to solve for each of the two resulting sums in terms of the E_{2m+2}, and simplifying.

To establish (2.95), consider the elementary identity

$$\sum_{r=1}^{\infty} g(2r-1) = \sum_{r=1}^{\infty} g(r) - \sum_{r=1}^{\infty} g(2r), \tag{2.99}$$

with $g(r)$ determined by (2.83). Apply the trivial identity

$$\frac{x}{1-x^2} = \frac{x}{1-x} - \frac{x^2}{1-x^2} \tag{2.100}$$

termwise to the two resulting sums. Finally, use (2.91) or (2.92) to solve for each of the next four sums in terms of the E_{2m}, and simplify. For the case $m = 1$, note that $(-3/\pi y) + (9/2\pi y) + (-6/4\pi y) = 0$.

Equation (2.96) is an immediate consequence of applying (2.100) termwise to the sum for $D_{2m+1}(q)$ in (2.84), using (2.92) to solve for each of the two resulting sums interms of the E_{2m+2}, and simplifying. □

The $m = 1, 2, 3, 4$ cases of the identities in Lemma 2.6 either appear or are implicit in [21]. The $m = 1, 2, 3$ cases of (2.93) are given by [21, Entries 14(i)–(iii) and their proofs, pp. 129–131], and the $m = 4$ case is implicit in [21, Entry 14(iv), p. 130]. The $m = 1, 2$ cases of (2.94) are given by [21, Entries 14(v)–(vi) and their proofs, pp. 130–131], and the $m = 3, 4$ cases are implicit in [21, Entries 14(vii)–(viii), p. 130]. The $m = 1, 2, 3$ cases of (2.95) are given by [21, Entries 15(ix)–(xi) and their proofs, pp. 132–133], and the $m = 4$ case is implicit in [21, Entry 15(xii), p. 133]. The $m = 1, 2$ cases of (2.96) are given by

[21, Entries 15(i)–(ii) and their proofs, pp. 132–133], and the $m = 3, 4$ cases are implicit in [21, Entries 15(iii)–(iv), p. 132].

Equation (2.98) is utilized in series manipulations in [21, pp. 226, 260, 383], and (2.100) is applied in the proof of [21, Entry 15(ix), pp. 132–133]. See also the comment just after Eq. (5.180).

The Lambert series $U_{2m-1}(q)$, $G_{2m+1}(q)$, $C_{2m-1}(q)$, and $D_{2m+1}(q)$ in Lemma 2.6 appear in the sums of squares and sums of triangles identities in Theorems 5.3, 5.4, 5.5, 5.6, 5.11, and 5.12.

3. Continued fraction expansions

In this section we derive associated continued fraction and regular C-fraction expansions of the Laplace transform and formal Laplace transform of various ratios of Jacobi elliptic functions. We first survey Rogers' [207] integration-by-parts proof of the associated continued fraction expansions of the Laplace transform of sn, cn, dn, and sn^2. We also provide a similar proof of the associated continued fraction expansions of the Laplace transform of sn cn and sn dn. The sn cn case was first obtained by Ismail and Masson in [111] using a more refined integration-by-parts analysis. We next recall Rogers' application of Landen's transformation [250, p. 507], [134, Eq. (3.9.15), (3.9.16), (3.9.17), pp. 78–79] to his result for sn to obtain the associated continued fraction expansion of the Laplace transform of sn cn/dn. We apply a similar modular transformation technique to obtain serveral other continued fraction expansions. Our formal Laplace transform continued fraction expansions are a consequence of those for sn, cn, dn, sn^2, sn cn, sn dn, combined with Heilermann's [103, 104] correspondence between formal power series and either associated continued fractions or regular C-fractions (see also [119, Theorem 7.14, pp. 244–246; Theorem 7.2, pp. 223–224]), and modular transformations.

Following Jones and Thron in [119, pp. 18–19], Lorentzen and Waadeland in [149, pp. 5–8], and Berndt in [20, pp. 104–105], we adopt the following notation for continued fractions:

$$\mathop{\mathbf{K}}_{n=1}^{\infty} \frac{a_n}{b_n} := \cfrac{a_1}{b_1 + \cfrac{a_2}{b_2 + \cfrac{a_3}{b_3 + \ddots}}} = \frac{a_1}{b_1} \, \frac{}{+} \, \frac{a_2}{b_2} \, \frac{}{+} \, \frac{a_3}{b_3} \, \frac{}{+} \cdots \tag{3.1}$$

The types of continued fractions that we need in this paper are summarized in the following definition.

Definition 3.1. Let $\{\alpha_\nu\}_{\nu=1}^{\infty}$, $\{\beta_\nu\}_{\nu=1}^{\infty}$, and $\{\gamma_\nu\}_{\nu=1}^{\infty}$ be sequences in $\mathbb{C}^{\times}$ with $\alpha_\nu\gamma_\nu \neq 0$, and let w and ξ be indeterminate. The "corresponding type" continued fraction or regular C-fraction is given by

$$1 + \mathop{\mathbf{K}}_{n=1}^{\infty} \frac{\gamma_n w}{1}, \quad \gamma_n \neq 0. \tag{3.2}$$

The associated continued fraction is given by

$$1+\frac{\alpha_1 w}{1+\beta_1 w+}\mathop{\mathbf{K}}_{n=2}^{\infty}\frac{-\alpha_n w^2}{1+\beta_n w},\quad \alpha_n\neq 0. \tag{3.3}$$

The Jacobi continued fraction or J-fraction is given by

$$\frac{\alpha_1}{\beta_1+\xi+}\mathop{\mathbf{K}}_{n=2}^{\infty}\frac{-\alpha_n}{\beta_n+\xi},\quad \alpha_n\neq 0. \tag{3.4}$$

The following sources have been used for Definition 3.1. For regular C-fractions, see [119, Eq. (7.1.1), p. 221], [149, pp. 252–253], [185, p. 304], [248, Eq. (99.2), p. 399; Eq. (54.2), p. 208]. The associated continued fraction appears in [119, Eq. (7.2.1), p. 241], [185, pp. 322, 324; Eq. (8), p. 376], [248, Eq. (54.1), p. 208]. Finally, the J-fraction can be found in [119, Eq. (7.2.35), p. 249], [149, p. 346], [185, Eq. (9), p. 376; Eq. (1), p. 390], [248, Eq. (23.8), p. 103; Eq. (51.1), p. 196]. J-fractions are so named because the related quadratic form has long been called a J-form. For all three types of continued fractions, see [119, pp. 128–129; Appendix A. pp. 386–394].

It is well-known (see [119, p. 129], [206, p. 74]) that the even part [119, Eq. (2.4.24), p. 42] of a regular C-fraction is an associated continued fraction. Moreover, [119, p. 249], if in the associated continued fraction (3.3) we let $w=1/\xi$, omit the initial term 1 and make an equivalence transformation, we obtain the J-fraction in (3.4). See [119, pp. 249–256] for the connection between J-fractions and orthogonal polynomials. Most of our work is with associated continued fractions, some is with regular C-fractions, and we just mention J-fractions to make the connection with orthogonal polynomials.

Several recent authors used the term J-fraction for what is really an associated continued fraction. They include: Flajolet [69, p. 130], [70, p. 146], Goulden and Jackson [92, Definition 5.2.1, p. 291], and Zeng [257, p. 374].

We now have the following associated continued fraction expansions of Laplace transforms of various ratios of Jacobi elliptic functions.

Theorem 3.2. *Let the Jacobi elliptic functions have modulus k, and let $k' := \sqrt{1-k^2}$. We then have the associated continued fraction expansions:*

$$\int_0^\infty \operatorname{sn} u\, e^{-u/x}\,du = \frac{x^2}{1+(1+k^2)x^2+}\mathop{\mathbf{K}}_{n=2}^{\infty}\frac{-(2n-1)(2n-2)^2(2n-3)k^2x^4}{1+(2n-1)^2(1+k^2)x^2} \tag{3.5}$$

$$\int_0^\infty \operatorname{cn} u\, e^{-u/x}\,du = \frac{x}{1+x^2+}\mathop{\mathbf{K}}_{n=2}^{\infty}\frac{-(2n-2)^2(2n-3)^2k^2x^4}{1+((2n-1)^2+(2n-2)^2k^2)x^2} \tag{3.6}$$

$$\int_0^\infty \operatorname{dn} u\, e^{-u/x}\,du = \frac{x}{1+k^2x^2+}\mathop{\mathbf{K}}_{n=2}^{\infty}\frac{-(2n-2)^2(2n-3)^2k^2x^4}{1+((2n-1)^2k^2+(2n-2)^2)x^2} \tag{3.7}$$

$$\int_0^\infty \operatorname{sn} u \operatorname{cn} u\, e^{-u/x}\,du = \frac{x^2}{1+(4+k^2)x^2+}\mathop{\mathbf{K}}_{n=2}^{\infty}\frac{-(2n-2)^2(2n-1)^2k^2x^4}{1+((2n)^2+(2n-1)^2k^2)x^2} \tag{3.8}$$

$$\int_0^\infty \operatorname{sn} u \operatorname{dn} u \, e^{-u/x} \, du = \frac{x^2}{1+(1+4k^2)x^2} + \mathop{\mathbf{K}}_{n=2}^{\infty} \frac{-(2n-2)^2(2n-1)^2k^2x^4}{1+((2n-1)^2+(2n)^2k^2)x^2} \quad (3.9)$$

$$\int_0^\infty \operatorname{sd} u \, e^{-u/x} \, du = \frac{x^2}{1+(1-2k^2)x^2} + \mathop{\mathbf{K}}_{n=2}^{\infty} \frac{(2n-1)(2n-2)^2(2n-3)(kk')^2x^4}{1+(2n-1)^2(1-2k^2)x^2} \quad (3.10)$$

$$\int_0^\infty \operatorname{cd} u \, e^{-u/x} \, du = \frac{x}{1+k'^2x^2} + \mathop{\mathbf{K}}_{n=2}^{\infty} \frac{(2n-2)^2(2n-3)^2(kk')^2x^4}{1+((2n-1)^2k'^2-(2n-2)^2k^2)x^2} \quad (3.11)$$

$$\int_0^\infty \operatorname{nd} u \, e^{-u/x} \, du = \frac{x}{1-k^2x^2} + \mathop{\mathbf{K}}_{n=2}^{\infty} \frac{(2n-2)^2(2n-3)^2(kk')^2x^4}{1+((2n-2)^2k'^2-(2n-1)^2k^2)x^2} \quad (3.12)$$

$$\int_0^\infty \frac{\operatorname{sn} u \operatorname{cn} u}{\operatorname{dn}^2 u} e^{-u/x} \, du = \frac{x^2}{1+(4-5k^2)x^2} + \mathop{\mathbf{K}}_{n=2}^{\infty} \frac{(2n-2)^2(2n-1)^2(kk')^2x^4}{1+((2n)^2k'^2-(2n-1)^2k^2)x^2} \quad (3.13)$$

$$\int_0^\infty \frac{\operatorname{sn} u}{\operatorname{dn}^2 u} e^{-u/x} \, du = \frac{x^2}{1+(1-5k^2)x^2} + \mathop{\mathbf{K}}_{n=2}^{\infty} \frac{(2n-2)^2(2n-1)^2(kk')^2x^4}{1+((2n-1)^2k'^2-(2n)^2k^2)x^2} \quad (3.14)$$

$$\int_0^\infty \operatorname{sn}^2 u \, e^{-u/x} \, du = \frac{2x^3}{1+4(1+k^2)x^2} + \mathop{\mathbf{K}}_{n=2}^{\infty} \frac{-(2n)(2n-1)^2(2n-2)k^2x^4}{1+(2n)^2(1+k^2)x^2} \quad (3.15)$$

$$\int_0^\infty \operatorname{sd}^2 u \, e^{-u/x} \, du = \frac{2x^3}{1+4(1-2k^2)x^2} + \mathop{\mathbf{K}}_{n=2}^{\infty} \frac{(2n)(2n-1)^2(2n-2)(kk')^2x^4}{1+(2n)^2(1-2k^2)x^2} \quad (3.16)$$

$$\int_0^\infty \frac{\operatorname{sn} u \operatorname{cn} u}{\operatorname{dn} u} e^{-u/x} \, du = \frac{x^2}{1+(4-2k^2)x^2} + \mathop{\mathbf{K}}_{n=2}^{\infty} \frac{-(2n-1)(2n-2)^2(2n-3)k^4x^4}{1+(2n-1)^2(4-2k^2)x^2} \quad (3.17)$$

$$\int_0^\infty \frac{\operatorname{sn}^2 u \operatorname{cn}^2 u}{\operatorname{dn}^2 u} e^{-u/x} \, du = \frac{2x^3}{1+4(4-2k^2)x^2} + \mathop{\mathbf{K}}_{n=2}^{\infty} \frac{-(2n)(2n-1)^2(2n-2)k^4x^4}{1+(2n)^2(4-2k^2)x^2} \quad (3.18)$$

$$\int_0^\infty \frac{1-k\operatorname{sn}^2 u}{1+k\operatorname{sn}^2 u} e^{-u/x} \, du = \frac{x}{1+4kx^2} + \mathop{\mathbf{K}}_{n=2}^{\infty} \frac{-4(2n-2)^2(2n-3)^2k(1+k)^2x^4}{1+((2n-1)^2 4k+(2n-2)^2(1+k)^2)x^2} \quad (3.19)$$

$$\int_0^\infty \frac{\operatorname{sn} u}{1+k\operatorname{sn}^2 u} e^{-u/x} \, du = \frac{x^2}{1+(1+6k+k^2)x^2} + \mathop{\mathbf{K}}_{n=2}^{\infty} \frac{-4(2n-1)(2n-2)^2(2n-3)k(1+k)^2x^4}{1+(2n-1)^2(1+6k+k^2)x^2} \quad (3.20)$$

$$\int_0^\infty \frac{\operatorname{cn} u \operatorname{dn} u}{1+k\operatorname{sn}^2 u} e^{-u/x}\,du$$
$$= \frac{x}{1+(1+k)^2x^2 + \underset{n=2}{\overset{\infty}{\mathrm{K}}} \frac{-4(2n-2)^2(2n-3)^2k(1+k)^2x^4}{1+((2n-1)^2(1+k)^2+(2n-2)^2 4k)x^2}} \tag{3.21}$$

Proof: We first use Rogers' [207] integration-by-parts argument to establish Eq. (3.5), (3.6), (3.7), and (3.15).

We begin by defining S_n, C_n, D_n, with $n = 0, 1, 2, \ldots$, by

$$S_n := \int_0^\infty \operatorname{sn}^n(u,k)\, e^{-u/x}\,du, \tag{3.22}$$
$$C_n := \int_0^\infty \operatorname{sn}^n(u,k)\operatorname{cn}(u,k)\, e^{-u/x}\,du, \tag{3.23}$$
$$D_n := \int_0^\infty \operatorname{sn}^n(u,k)\operatorname{dn}(u,k)\, e^{-u/x}\,du. \tag{3.24}$$

We integrate by parts and use various differentiation formulas and Pythagorean theorems for Jacobi elliptic functions to show that:

$$S_1 = x^2 - x^2(1+k^2)S_1 + 2k^2x^2S_3, \tag{3.25}$$

and

$$S_n = n(n-1)x^2S_{n-2} - n^2(1+k^2)x^2S_n + n(n+1)k^2x^2S_{n+2}, \tag{3.26}$$

for $n = 2, 3, 4 \ldots$.

Solving for S_1 and S_n/S_{n-2} in just the right way, we obtain:

$$S_1 = \frac{x^2}{1+(1+k^2)x^2 - 2k^2x^2S_3/S_1}, \tag{3.27}$$
$$S_n/S_{n-2} = \frac{n(n-1)x^2}{1+n^2(1+k^2)x^2 - n(n+1)k^2x^2S_{n+2}/S_n}, \tag{3.28}$$

for $n = 2, 3, 4 \ldots$.

After iterating and simplifying, we obtain the continued fraction expansion in (3.5).

Using machinery from the proof of (3.5), and noting that $S_0 = x$, we have

$$S_2 = \frac{2x^3}{1+4(1+k^2)x^2 - 6k^2x^2S_4/S_2}, \tag{3.29}$$

with S_n/S_{n-2} as in (3.28). Again, we iterate and simplify to obtain the continued fraction expansion in (3.15).

Proceeding as in the proof of (3.5), we obtain:

$$C_0 = x - x^2C_0 + 2k^2x^2C_2, \tag{3.30}$$

and

$$C_n = n(n-1)x^2C_{n-2} - ((n+1)^2 + k^2n^2)x^2C_n + (n+1)(n+2)k^2x^2C_{n+2}, \quad (3.31)$$

for $n = 2, 3, 4 \ldots$.
Solving for C_0 and C_n/C_{n-2}, we obtain:

$$C_0 = \frac{x}{1 + x^2 - 2k^2x^2C_2/C_0}, \quad (3.32)$$

$$C_n/C_{n-2} = \frac{n(n-1)x^2}{1 + ((n+1)^2 + k^2n^2)x^2 - (n+1)(n+2)k^2x^2C_{n+2}/C_n}, \quad (3.33)$$

for $n = 2, 3, 4 \ldots$.
After iterating and simplifying, we obtain the continued fraction expansion in (3.6).

Next, proceeding as in the proof of (3.6), we obtain:

$$D_0 = x - k^2x^2D_0 + 2k^2x^2D_2, \quad (3.34)$$

and

$$D_n = n(n-1)x^2D_{n-2} - (n^2 + (n+1)^2k^2)x^2D_n + (n+1)(n+2)k^2x^2D_{n+2}, \quad (3.35)$$

for $n = 2, 3, 4 \ldots$.
Solving for D_0 and D_n/D_{n-2}, we obtain:

$$D_0 = \frac{x}{1 + k^2x^2 - 2k^2x^2D_2/D_0}, \quad (3.36)$$

$$D_n/D_{n-2} = \frac{n(n-1)x^2}{1 + (n^2 + (n+1)^2k^2)x^2 - (n+1)(n+2)k^2x^2D_{n+2}/D_n}, \quad (3.37)$$

for $n = 2, 3, 4 \ldots$.
After iterating and simplifying, we obtain the continued fraction expansion in (3.7).

We now use a slight variation of Rogers' integration by parts arguments to obtain Eqs. (3.8) and (3.9). The key idea is to utilize just one integration by parts at a time, instead of two successive integrations, to derive formulas for C_1 and D_1 that are analogous to those for C_0 and D_0 in (3.32) and (3.36), respectively. The rest of the analysis then iterates suitable cases of (3.33) and (3.37) as before.

Starting with the C_n and D_n in (3.23) and (3.24), it follows that one integration by parts and the Pythagorean relation for $\mathrm{cn}^2(u, k)$ or $\mathrm{dn}^2(u, k)$, respectively, gives the identities

$$C_n = nxD_{n-1} - (n+1)xD_{n+1}, \quad (3.38)$$

$$D_n = nxC_{n-1} - (n+1)k^2xC_{n+1}, \quad (3.39)$$

for $n = 1, 2, 3 \ldots$.

We first obtain our formula for C_1. Setting $n = 1$ in (3.38) gives

$$C_1 = xD_0 - 2xD_2. \quad (3.40)$$

One integration by parts applied to D_0 gives

$$D_0 = x - xk^2C_1. \tag{3.41}$$

From the $n = 2$ case of (3.39) we have

$$D_2 = 2xC_1 - 3k^2xC_3. \tag{3.42}$$

Substituting (3.41) and (3.42) into (3.40) immediately gives

$$C_1 = x^2 - x^2(4+k^2)C_1 + 6k^2x^2C_3. \tag{3.43}$$

Solving for C_1 in (3.43) in just the right way yields

$$C_1 = \frac{x^2}{1+(4+k^2)x^2 - 6k^2x^2C_3/C_1}. \tag{3.44}$$

We next obtain the formula for D_1. Setting $n = 1$ in (3.39) gives

$$D_1 = xC_0 - 2k^2xC_2. \tag{3.45}$$

One integration by parts applied to C_0 gives

$$C_0 = x - xD_1. \tag{3.46}$$

From the $n = 2$ case of (3.38) we have

$$C_2 = 2xD_1 - 3xD_3. \tag{3.47}$$

Substituting (3.46) and (3.47) into (3.45) immediately gives

$$D_1 = x^2 - x^2(1+4k^2)D_1 + 6k^2x^2D_3. \tag{3.48}$$

Solving for D_1 in (3.48) in just the right way yields

$$D_1 = \frac{x^2}{1+(1+4k^2)x^2 - 6k^2x^2D_3/D_1}. \tag{3.49}$$

Note that substituting suitable cases of (3.39) into (3.38), or of (3.38) into (3.39) yields (3.31) and (3.35), respectively. These in turn lead to (3.33) and (3.37).

After iterating (3.44) and (3.33), and simplifying, we obtain the continued fraction expansion in (3.8). Similarly, iterating (3.49) and (3.37) leads to (3.9).

A short alternate proof of (3.8) and (3.9) is included just after the proof of Theorem 3.11 near the end of this section.

We now use Rogers' [207] modular transformation technique to establish (3.10)–(3.14) and (3.16)–(3.21).

Rogers [207] deduced (3.17) and (3.18) by applying Landen's transformation [250, p. 507] in the form

$$\frac{\mathrm{sn}(u,k)\,\mathrm{cn}(u,k)}{\mathrm{dn}(u,k)} = \frac{1}{1+k'}\,\mathrm{sn}\left((1+k')u, \frac{1-k'}{1+k'}\right) \tag{3.50}$$

to the integrands in (3.17) and (3.18), utilizing the change of variables $v = (1+k')u$, appealing to (3.5) and (3.15), and then simplifying. Here, we use the fact that

$$k_1^2(1+k')^4 = k^4 \quad \text{and} \quad (1+k_1^2)(1+k')^2 = 2(2-k^2), \tag{3.51}$$

where $k' := \sqrt{1-k^2}$ and $k_1 := (1-k')/(1+k')$.

In order to establish (3.10), (3.11), (3.12), and (3.16) consider the modular transformations [134, Example 33, p. 90; and Eq. (9.3.9), pp. 249–250], [250, Section 22.421, p. 508] in

$$\mathrm{sd}(u,k) = \frac{1}{k'}\,\mathrm{sn}(k'u, ik/k'), \tag{3.52a}$$

$$\mathrm{cd}(u,k) = \mathrm{cn}(k'u, ik/k'), \tag{3.52b}$$

$$\mathrm{nd}(u,k) = \mathrm{dn}(k'u, ik/k'). \tag{3.52c}$$

Next, apply (3.52a), (3.52b), (3.52c) as needed to the integrands in (3.10), (3.11), (3.12), (3.13), (3.14), and (3.16). Change variables by $v = k'u$, appeal to (3.5), (3.6), (3.7), (3.8), (3.9), (3.15), and then simplify.

The continued fractions in (3.20), (3.21), and (3.19) are immediate consequences of applying the Gauß modular transformations [134, p. 80], [93, p. 915] in

$$\frac{\mathrm{sn}(u,k)}{1+k\,\mathrm{sn}^2(u,k)} = \frac{1}{1+k}\,\mathrm{sn}\left((1+k)u, \frac{2\sqrt{k}}{1+k}\right), \tag{3.53a}$$

$$\frac{\mathrm{cn}(u,k)\,\mathrm{dn}(u,k)}{1+k\,\mathrm{sn}^2(u,k)} = \mathrm{cn}\left((1+k)u, \frac{2\sqrt{k}}{1+k}\right), \tag{3.53b}$$

$$\frac{1-k\,\mathrm{sn}^2(u,k)}{1+k\,\mathrm{sn}^2(u,k)} = \mathrm{dn}\left((1+k)u, \frac{2\sqrt{k}}{1+k}\right), \tag{3.53c}$$

to the integrands in (3.20), (3.21), and (3.19), respectively, changing variables by $v = (1+k)u$, appealing to (3.5), (3.6), and (3.7), and then simplifying. □

The integration-by-parts proof of (3.5), (3.6), (3.7), and (3.15) first appeared in Rogers' [207, pp. 76–77]. A more recent discussion of these calculations can be found in [92, pp. 307–308, Example 5.2.8, pp. 517–519] and [71]. The sn cn case in (3.8) was first obtained by Ismail and Masson in [111] using a more refined integration-by-parts analysis. Stieltjes [218, 220] first derived (3.5), (3.6), (3.7), and (3.15) by his addition theorem for elliptic functions method. Rogers [207] also rediscovered the addition theorem techniques of Stieltjes. Elegant combinatorial applications of (3.6) and/or the addition theorem techniques of Rogers and Stieltjes are studied in [29, 30, 58, 59, 69–71, 97, 98, 196–200, 224, 256–258]. Rogers [207] was the first to derive the associated continued fraction expansions in

(3.17) and (3.18) by applying Landen's transformation [250, p. 507], [135, Eq. (3.9.15), (3.9.16), (3.9.17), pp. 78–79] to (3.5), and (3.15), respectively. Ramanujan was the first to obtain continued fraction expansions equivalent to (3.10) and (3.11), in which the integral is written as a hyperbolic series. In particular, Ramanujan had the associated continued fraction equivalent to (3.10), and the regular C-fraction equivalent to (3.107) below. Berndt [21, pp. 165–167] obtains Ramanujan's two results by applying the appropriate modular transformations to (3.5) and (3.6). The rest of the associated continued fraction expansions in Theorem 3.2 appear to be new. Subsequently, a simplified and more symmetrical approach to Theorem 3.2 appears in [49].

In order to derive our formal Laplace transform associated continued fraction expansions from the continued fractions in Theorem 3.2 we first need Heilermann's [103, 104] correspondence between formal power series and associated continued fractions. Furthermore, the derivation of our formal Laplace transform C-fraction expansions in (3.109) and (3.110) below requires Heilermann's [103, 104] correspondence between formal power series and C-fractions.

With any formal power series

$$L(w) = 1 + c_1 w + c_2 w^2 + c_3 w^3 + \cdots, \tag{3.54}$$

where $\{c_\nu\}_{\nu=1}^{\infty}$ is a sequence in $\mathbb{C}^\times$, we associate the two sequences of determinants of $n \times n$ square matrices given by the following definition.

Definition 3.3. Let $\{c_\nu\}_{\nu=1}^{\infty}$ be a sequence in $\mathbb{C}^\times$, and let $m, n = 1, 2, 3, \ldots$. We take $H_n^{(m)}$ and χ_n to be the determinants of $n \times n$ square matrices

$$H_n^{(m)} \equiv H_n^{(m)}(\{c_\nu\}) := \det \begin{pmatrix} c_m & c_{m+1} & \cdots & c_{m+n-2} & c_{m+n-1} \\ c_{m+1} & c_{m+2} & \cdots & c_{m+n-1} & c_{m+n} \\ \vdots & \vdots & \ddots & \vdots & \vdots \\ c_{m+n-1} & c_{m+n} & \cdots & c_{m+2n-3} & c_{m+2n-2} \end{pmatrix}, \tag{3.55}$$

$$\chi_n \equiv \chi_n(\{c_\nu\}) := \det \begin{pmatrix} c_1 & c_2 & \cdots & c_{n-1} & c_{n+1} \\ c_2 & c_3 & \cdots & c_n & c_{n+2} \\ \vdots & \vdots & \ddots & \vdots & \vdots \\ c_n & c_{n+1} & \cdots & c_{2n-2} & c_{2n} \end{pmatrix} \tag{3.56}$$

The matrix for χ_n is obtained from the matrix for $H_{n+1}^{(1)}$ by deleting the next to last column and the last row. In particular, for $n = 1$ we have $H_1^{(1)} = c_1$, $H_1^{(2)} = c_2$, and $\chi_1 = c_2$.[1] We also have $H_0^{(n)} = 1$ and $\chi_0 = 0$. Note that $H_n^{(m)}(\{c_\nu\})$ is not the Hankel function in [6, p. 208].

The following theorem [103, 104], [119, Theorem 7.14, pp. 244–246] provides explicit necessary and sufficient conditions for expanding a formal power series into an associated continued fraction.

Theorem 3.4 (Heilermann). *If for a given formal power series $L(w)$ in* (3.54) *we have*

$$1+\sum_{m=1}^{\infty} c_m w^m = 1 + \frac{\alpha_1 w}{1+\beta_1 w +} \mathop{\mathbf{K}}_{n=2}^{\infty} \frac{-\alpha_n w^2}{1+\beta_n w}, \quad \alpha_n \neq 0, \tag{3.57}$$

then

$$H_n^{(1)}(\{c_\nu\}) \neq 0, \quad \textit{for } n = 1, 2, 3, \ldots, \tag{3.58}$$

where $H_n^{(1)}(\{c_\nu\})$ is the Hankel determinant in (3.55) *associated with $L(w)$. Moreover,*

$$\alpha_n = \frac{H_n^{(1)} H_{n-2}^{(1)}}{\left(H_{n-1}^{(1)}\right)^2}, \quad \textit{for } n = 1, 2, 3, \ldots, \quad \left(H_{-1}^{(1)} = H_0^{(1)} = 1\right),$$

and

$$\beta_n = \frac{\chi_{n-1}}{H_{n-1}^{(1)}} - \frac{\chi_n}{H_n^{(1)}}, \quad \textit{for } n = 1, 2, 3, \ldots, \quad (\chi_0 = 0, \ \chi_1 = c_2), \tag{3.59}$$

where χ_n is given by (3.56).

Conversely, suppose that (3.58) *holds. Then,* (3.57) *also holds with coefficients $\{\alpha_n\}$ and $\{\beta_n\}$ given by* (3.59)*. In addition, we have*

$$H_n^{(1)}(\{c_\nu\}) = \prod_{r=1}^{n} \alpha_r^{n+1-r}, \quad \textit{for } n = 1, 2, 3, \ldots, \tag{3.60}$$

and

$$\begin{aligned} \chi_n(\{c_\nu\}) &= -(\beta_1+\beta_2+\cdots+\beta_n)\, H_n^{(1)}(\{c_\nu\}) \\ &= -(\beta_1+\beta_2+\cdots+\beta_n) \prod_{r=1}^{n} \alpha_r^{n+1-r}, \quad \textit{for } n = 1, 2, 3, \ldots. \end{aligned} \tag{3.61}$$

The following theorem [103, 104], [119, Theorem 7.2, pp. 223–224] provides explicit necessary and sufficient conditions for expanding a formal power series into a regular C-fraction.

Theorem 3.5 (Heilermann). *If for a given formal power series $L(w)$ in* (3.54) *we have*

$$1+\sum_{m=1}^{\infty} c_m w^m = 1 + \mathop{\mathbf{K}}_{n=1}^{\infty} \frac{\gamma_n w}{1}, \quad \gamma_n \neq 0, \tag{3.62}$$

then

$$H_n^{(1)}(\{c_\nu\}) \neq 0 \quad \textit{and} \quad H_n^{(2)}(\{c_\nu\}) \neq 0, \quad \textit{for } n = 1, 2, 3, \ldots, \tag{3.63}$$

where $H_n^{(1)}(\{c_\nu\})$ and $H_n^{(2)}(\{c_\nu\})$ are the Hankel determinants in (3.55) associated with $L(w)$. Moreover,

$$\gamma_{2m} = -\frac{H_{m-1}^{(1)}H_m^{(2)}}{H_m^{(1)}H_{m-1}^{(2)}}, \quad \text{for } m = 1,2,3,\ldots, \quad (H_0^{(1)} = H_0^{(2)} = 1),$$

$$\gamma_1 = H_1^{(1)}; \; \gamma_{2m+1} = -\frac{H_{m+1}^{(1)}H_{m-1}^{(2)}}{H_m^{(1)}H_m^{(2)}}, \quad \text{for } m = 1,2,3,\ldots, \quad (H_0^{(2)} = 1). \tag{3.64}$$

Conversely, suppose that (3.63) *holds. Then,* (3.62) *also holds with coefficients $\{\gamma_n\}$ given by* (3.64). *In addition, we have*

$$\begin{aligned} H_n^{(2)}(\{c_\nu\}) &= (-1)^n H_n^{(1)}(\{c_\nu\}) \prod_{r=1}^{n} \gamma_{2r} \\ &= (-1)^n H_{n+1}^{(1)}(\{c_\nu\}) \prod_{r=0}^{n} \gamma_{2r+1}^{-1}, \quad \text{for } n = 1,2,3,\ldots. \end{aligned} \tag{3.65}$$

In our applications of Theorems 3.4 and 3.5 in this paper, we find the following elementary lemma very useful.

Lemma 3.6. *Let $\{c_\nu\}_{\nu=1}^{\infty}$ be a sequence in $\mathbb{C}^\times$, let $n = 1,2,3,\ldots$, and take $H_n^{(m)}$ and χ_n as in Definition 3.3. If x is a constant we then have*

$$H_n^{(1)}(\{x^\nu c_\nu\}) = x^{n^2} H_n^{(1)}(\{c_\nu\}), \tag{3.66}$$

$$H_n^{(1)}(\{x^{\nu-1}c_{\nu-1}\}) = x^{n(n-1)} H_n^{(1)}(\{c_{\nu-1}\}) = x^{2\binom{n}{2}}\, H_n^{(1)}(\{c_{\nu-1}\}), \tag{3.67}$$

$$\chi_n(\{x^\nu c_\nu\}) = x^{1+n^2}\chi_n(\{c_\nu\}), \tag{3.68}$$

$$\chi_n(\{x^{\nu-1}c_{\nu-1}\}) = x^{1+n(n-1)}\chi_n(\{c_{\nu-1}\}) = x^{1+2\binom{n}{2}}\,\chi_n(\{c_{\nu-1}\}). \tag{3.69}$$

$$H_n^{(2)}(\{x^\nu c_\nu\}) = x^{n(n+1)} H_n^{(2)}(\{c_\nu\}), \tag{3.70}$$

$$H_n^{(2)}(\{x^{\nu-1}c_{\nu-1}\}) = x^{n^2} H_n^{(2)}(\{c_{\nu-1}\}). \tag{3.71}$$

The relations in (3.66)–(3.71) follow immediately by first factoring suitable powers of x from the rows, then the columns.

Applying (3.66) and (3.68) to (3.58) and (3.59), it is not difficult to see that we have the following lemma.

Lemma 3.7. *Suppose that the associated continued fraction expansion in* (3.57) *holds, and that A and B are nonzero constants. We then have*

$$1 + \sum_{m=1}^{\infty} AB^m c_m w^m = 1 + \frac{AB\alpha_1 w}{1 + B\beta_1 w +} \mathop{\mathbf{K}}_{n=2}^{\infty} \frac{-B^2\alpha_n w^2}{1 + B\beta_n w}, \tag{3.72}$$

where $\{\alpha_n\}$ and $\{\beta_n\}$ are given by (3.59). *That is, if $L(w)$ is the formal power series in* (3.54), *and* (3.57) *holds, then* (3.72) *gives the associated continued fraction expansion for $1 + A(L(Bw) - 1)$.*

Proof: The associated continued fraction expansion in (3.72) is an immediate consequence of assuming (3.57)–(3.59), and then using (3.66) and (3.68) to simplify (3.58) and (3.59) where the sequence $\{c_\nu\}_{\nu=1}^{\infty}$ is replaced by $\{AB^\nu c_\nu\}_{\nu=1}^{\infty}$. Just note that

$$H_n^{(1)}(\{AB^\nu c_\nu\}) = A^n B^{n^2} H_n^{(1)}(\{c_\nu\}), \tag{3.73}$$

$$\chi_n(\{AB^\nu c_\nu\}) = A^n B^{1+n^2} \chi_n(\{c_\nu\}), \tag{3.74}$$

and then substitute into the right-hand sides of (3.59). We find that α_1, α_n, and β_n become $AB\alpha_1$, $B^2\alpha_n$, and $B\beta_n$, respectively. We used the necessary and then sufficient conditions in Theorem 3.4. □

Next, applying (3.70) to (3.63) and (3.64), it is not difficult to see that we have the following lemma.

Lemma 3.8. *Suppose that the regular C-fraction fraction expansion in* (3.62) *holds, and that A and B are nonzero constants. We then have*

$$1+\sum_{m=1}^{\infty} AB^m c_m w^m = 1+\frac{AB\gamma_1 w}{1} \underset{+}{} \mathop{\mathbf{K}}_{n=2}^{\infty} \frac{B\gamma_n w}{1}, \tag{3.75}$$

where $\{\gamma_n\}$ *is given by* (3.64). *That is, if* $L(w)$ *is the formal power series in* (3.54), *and* (3.62) *holds, then* (3.75) *gives the regular C-fraction expansion for* $1 + A(L(Bw) - 1)$.

Proof: The regular C-fraction expansion in (3.75) is an immediate consequence of assuming (3.62)–(3.64), and then using (3.66) and (3.70) to simplify (3.63) and (3.64) where the sequence $\{c_\nu\}_{\nu=1}^{\infty}$ is replaced by $\{AB^\nu c_\nu\}_{\nu=1}^{\infty}$. Just note that

$$H_n^{(1)}(\{AB^\nu c_\nu\}) = A^n B^{n^2} H_n^{(1)}(\{c_\nu\}), \tag{3.76}$$

$$H_n^{(2)}(\{AB^\nu c_\nu\}) = A^n B^{n(n+1)} H_n^{(2)}(\{c_\nu\}), \tag{3.77}$$

and then substitute into the right-hand sides of (3.64). We find that γ_1, γ_{2m+1}, and γ_{2m} become $AB\gamma_1$, $B\gamma_{2m+1}$, and $B\gamma_{2m}$, respectively. We used the necessary and then sufficient conditions in Theorem 3.5. □

In order to apply Lemmas 3.7 and 3.8 to Theorem 3.2 and the first part of Theorem 3.11 to obtain additional continued fraction expansions we first need the formal Laplace transform in the following definition.

Definition 3.9. Given a Maclaurin series expansion

$$f(u) = \sum_{m=0}^{\infty} \frac{a_m u^m}{m!}, \tag{3.78}$$

we obtain a formal Laplace transform by integrating term by term:

$$\mathcal{L}(f, x^{-1}) := \sum_{m=0}^{\infty} \frac{a_m}{m!} \int_0^{\infty} u^m e^{-u/x}\, du = \sum_{m=0}^{\infty} a_m x^{m+1}. \tag{3.79}$$

We now have the following theorem.

Theorem 3.10. *Let the Jacobi elliptic functions have modulus k, and let $k' := \sqrt{1-k^2}$. Furthermore, take the formal Laplace transform in Definition 3.9 of the indicated Maclaurin series expansions from Definition 2.3. We then have the associated continued fraction expansions:*

$$\int_0^\infty \operatorname{sc} u\, e^{-u/x}\, du = \frac{x^2}{1+(k^2-2)x^2} + \mathop{\mathbf{K}}_{n=2}^{\infty} \frac{-(2n-1)(2n-2)^2(2n-3)k'^2x^4}{1+(2n-1)^2(k^2-2)x^2} \tag{3.80}$$

$$\int_0^\infty \operatorname{dc} u\, e^{-u/x}\, du = \frac{x}{1-k'^2x^2} + \mathop{\mathbf{K}}_{n=2}^{\infty} \frac{-(2n-2)^2(2n-3)^2k'^2x^4}{1-((2n-1)^2k'^2+(2n-2)^2)x^2} \tag{3.81}$$

$$\int_0^\infty \operatorname{nc} u\, e^{-u/x}\, du = \frac{x}{1-x^2} + \mathop{\mathbf{K}}_{n=2}^{\infty} \frac{-(2n-2)^2(2n-3)^2k'^2x^4}{1-((2n-1)^2+(2n-2)^2k'^2)x^2} \tag{3.82}$$

$$\int_0^\infty \operatorname{sc}^2 u\, e^{-u/x}\, du = \frac{2x^3}{1+4(k^2-2)x^2} + \mathop{\mathbf{K}}_{n=2}^{\infty} \frac{-(2n)(2n-1)^2(2n-2)k'^2x^4}{1+(2n)^2(k^2-2)x^2} \tag{3.83}$$

$$\int_0^\infty \frac{\operatorname{sn} u \operatorname{dn} u}{\operatorname{cn}^2 u} e^{-u/x}\, du = \frac{x^2}{1-(1+4k'^2)x^2} + \mathop{\mathbf{K}}_{n=2}^{\infty} \frac{-(2n-2)^2(2n-1)^2k'^2x^4}{1-((2n-1)^2+(2n)^2k'^2)x^2} \tag{3.84}$$

$$\int_0^\infty \frac{\operatorname{sn} u}{\operatorname{cn}^2 u} e^{-u/x}\, du = \frac{x^2}{1-(4+k'^2)x^2} + \mathop{\mathbf{K}}_{n=2}^{\infty} \frac{-(2n-2)^2(2n-1)^2k'^2x^4}{1-((2n)^2+(2n-1)^2k'^2)x^2} \tag{3.85}$$

$$\int_0^\infty \frac{\operatorname{sn} u \operatorname{dn} u}{\operatorname{cn} u} e^{-u/x}\, du = \frac{x^2}{1+2(2k^2-1)x^2} + \mathop{\mathbf{K}}_{n=2}^{\infty} \frac{-(2n-1)(2n-2)^2(2n-3)x^4}{1+2(2n-1)^2(2k^2-1)x^2} \tag{3.86}$$

$$\int_0^\infty \frac{\operatorname{sn} u}{\operatorname{cn} u \operatorname{dn} u} e^{-u/x}\, du = \frac{x^2}{1-(2+2k^2)x^2} + \mathop{\mathbf{K}}_{n=2}^{\infty} \frac{-(2n-1)(2n-2)^2(2n-3)k'^4x^4}{1-(2n-1)^2(2+2k^2)x^2} \tag{3.87}$$

$$\int_0^\infty \frac{\operatorname{sn}^2 u \operatorname{dn}^2 u}{\operatorname{cn}^2 u} e^{-u/x}\, du = \frac{2x^3}{1+8(2k^2-1)x^2} + \mathop{\mathbf{K}}_{n=2}^{\infty} \frac{-(2n)(2n-1)^2(2n-2)x^4}{1+2(2n)^2(2k^2-1)x^2} \tag{3.88}$$

$$\int_0^\infty \frac{\operatorname{sn}^2 u}{\operatorname{cn}^2 u \operatorname{dn}^2 u} e^{-u/x}\, du = \frac{2x^3}{1-4(2+2k^2)x^2} + \mathop{\mathbf{K}}_{n=2}^{\infty} \frac{-(2n)(2n-1)^2(2n-2)k'^4x^4}{1-(2n)^2(2+2k^2)x^2} \tag{3.89}$$

Proof: We start with a modular transformation

$$f(u,k) = Ag(Bu,k_1), \tag{3.90}$$

where A and B are nonzero constants, k_1 is a function of the modulus k, and f and g are the quotients of Jacobi elliptic functions in the integrands of the (formal) Laplace transforms in Theorems 3.10 and 3.2, respectively.

Let the Maclaurin series expansions of $f(u,k)$ and $g(u,k)$ in Definition 2.3 be given by one of

$$f(u,k) = \sum_{m=0}^{\infty} f_m(k^2)\frac{u^{2m}}{(2m)!} \quad \text{and} \quad g(u,k) = \sum_{m=0}^{\infty} g_m(k^2)\frac{u^{2m}}{(2m)!}, \tag{3.91}$$

$$f(u,k) = \sum_{m=1}^{\infty} f_m(k^2)\frac{u^{2m-1}}{(2m-1)!} \quad \text{and} \quad g(u,k) = \sum_{m=1}^{\infty} g_m(k^2)\frac{u^{2m-1}}{(2m-1)!}, \tag{3.92}$$

$$f(u,k) = \sum_{m=1}^{\infty} f_m(k^2)\frac{u^{2m}}{(2m)!} \quad \text{and} \quad g(u,k) = \sum_{m=1}^{\infty} g_m(k^2)\frac{u^{2m}}{(2m)!}. \tag{3.93}$$

Substitute (3.91), (3.92), or (3.93) into (3.90), take the formal Laplace transform of both sides, multiply both sides by x, 1, or x^{-1}, respectively, and then add 1 to both sides. We obtain

$$1+\sum_{m=1}^{\infty} \bar{f}_m(k^2)w^m = 1+\sum_{m=1}^{\infty} CB^{2m}\,\bar{g}_m(k_1^2)w^m, \tag{3.94}$$

where $w = x^2$ and either $\bar{f}_m = f_{m-1}, \bar{g}_m = g_{m-1}, C = AB^{-2}$ in (3.91); $\bar{f}_m = f_m, \bar{g}_m = g_m$, $C = AB^{-1}$ in (3.92); and $\bar{f}_m = f_m, \bar{g}_m = g_m, C = A$ in (3.93).

The formulas for the derivatives (with respect to u) of $\operatorname{sn}(u,k)$, $\operatorname{cn}(u,k)$, and $\operatorname{dn}(u,k)$ are the same for $k > 1$ as they are for $k < 1$. Thus, we are able to use $g_{m-1}(k_1^2)$ and $g_m(k_1^2)$ in (3.94).

The associated continued fraction expansions in Theorem 3.10 are a direct consequence of applying Lemma 3.7 to the right hand side of the relations in (3.94), corresponding to suitable cases of (3.90), while keeping in mind the associated continued fraction expansions from Theorem 3.2 of the Laplace transform of the $g(u,k)$.

We complete the proof of Theorem 3.10 by writing down the necessary cases of (3.90).

For the associated continued fraction expansions (3.80), (3.81), and (3.82) consider Jacobi's imaginary modular transformations [134, Example 34, p. 91; and Eq. (9.4.2), p. 250], [250, Section 22.4, pp. 505–506] in

$$\operatorname{sc}(u,k) = -\frac{i}{k'}\operatorname{sn}(ik'u, 1/k'), \tag{3.95a}$$

$$\operatorname{dc}(u,k) = \operatorname{cn}(ik'u, 1/k'), \tag{3.95b}$$

$$\operatorname{nc}(u,k) = \operatorname{dn}(ik'u, 1/k'). \tag{3.95c}$$

For (3.83), we just need the square of (3.95a) in

$$\operatorname{sc}^2(u,k) = -\frac{1}{k'^2}\operatorname{sn}^2(ik'u, 1/k'). \tag{3.96}$$

For (3.84) and (3.85) we utilize the combinations of (3.95a–3.95c) given by

$$\frac{\operatorname{sn}(u,k)\operatorname{dn}(u,k)}{\operatorname{cn}^2(u,k)} = -\frac{i}{k'}\operatorname{sn}(ik'u,1/k')\operatorname{cn}(ik'u,1/k'), \tag{3.97}$$

$$\frac{\operatorname{sn}(u,k)}{\operatorname{cn}^2(u,k)} = -\frac{i}{k'}\operatorname{sn}(ik'u,1/k')\operatorname{dn}(ik'u,1/k'). \tag{3.98}$$

The associated continued fraction expansion in (3.87) depends on the modular transformation [134, Example 35(i), p. 91] in

$$\frac{\operatorname{sn}(u,k)}{\operatorname{cn}(u,k)\operatorname{dn}(u,k)} = -\frac{i}{1+k}\operatorname{sn}\left(i(1+k)u, \frac{1-k}{1+k}\right), \tag{3.99}$$

while (3.89) requires the square of (3.99) in

$$\frac{\operatorname{sn}^2(u,k)}{\operatorname{cn}^2(u,k)\operatorname{dn}^2(u,k)} = -\frac{1}{(1+k)^2}\operatorname{sn}^2\left(i(1+k)u, \frac{1-k}{1+k}\right). \tag{3.100}$$

To obtain (3.86) we observe from (3.95a)–(3.95c) that

$$\frac{\operatorname{sn}(u,k)\operatorname{dn}(u,k)}{\operatorname{cn}(u,k)} = \frac{\operatorname{sc}(u,k)\operatorname{dc}(u,k)}{\operatorname{nc}(u,k)} = -\frac{i}{k'}\frac{\operatorname{sn}(ik'u,1/k')\operatorname{cn}(ik'u,1/k')}{\operatorname{dn}(ik'u,1/k')}. \tag{3.101}$$

Finally, (3.88) follows from the square of (3.101) in

$$\frac{\operatorname{sn}^2(u,k)\operatorname{dn}^2(u,k)}{\operatorname{cn}^2(u,k)} = -\frac{1}{k'^2}\frac{\operatorname{sn}^2(ik'u,1/k')\operatorname{cn}^2(ik'u,1/k')}{\operatorname{dn}^2(ik'u,1/k')}. \tag{3.102}$$

□

As far as we know, the associated continued fraction expansions in Theorem 3.10 are new. It would be interesting to use suitable (modular) transformations to extend Theorems 3.2 and 3.10 to the setting of the more general types of continued fractions (associated with the Lamé equation) discussed in [46, pp. 28–31].

In order to obtain our regular C-fraction expansions we first recall from [119, p. 129] that the even part [119, Eq. (2.4.24), p. 42] of a regular C-fraction

$$\mathop{\mathbf{K}}_{n=1}^{\infty}\frac{\gamma_n w}{1}, \quad \gamma_n \neq 0, \tag{3.103}$$

is the associated continued fraction

$$\frac{\gamma_1 w}{1+\gamma_2 w} + \mathop{\mathbf{K}}_{n=1}^{\infty}\frac{-\gamma_{2n}\gamma_{2n+1}w^2}{1+(\gamma_{2n+1}+\gamma_{2n+2})w}. \tag{3.104}$$

To see why (3.103) and (3.104) are equal, note Rogers' [207, p. 74] observation that the $2m$-th convergent of (3.103) is identical with the m-th convergent of (3.104).

We now have the following regular C-fraction expansions of the Laplace transform and formal Laplace transform of various ratios of Jacobi elliptic functions.

Theorem 3.11. *Let the Jacobi elliptic functions have modulus k, and let $k' := \sqrt{1-k^2}$. Let $m = 1, 2, 3, \ldots$. We then have the regular C-fraction expansions*:

$$\int_0^\infty \operatorname{cn} u \, e^{-u/x} \, du = \frac{x}{1} + \mathop{\mathbf{K}}_{n=2}^{\infty} \frac{\gamma_n x^2}{1}, \tag{3.105a}$$

where

$$\gamma_{2m} = (2m-1)^2 \quad \textit{and} \quad \gamma_{2m+1} = (2m)^2 k^2. \tag{3.105b}$$

$$\int_0^\infty \operatorname{dn} u \, e^{-u/x} \, du = \frac{x}{1} + \mathop{\mathbf{K}}_{n=2}^{\infty} \frac{\gamma_n x^2}{1}, \tag{3.106a}$$

where

$$\gamma_{2m} = (2m-1)^2 k^2 \quad \textit{and} \quad \gamma_{2m+1} = (2m)^2. \tag{3.106b}$$

$$\int_0^\infty \operatorname{cd} u \, e^{-u/x} \, du = \frac{x}{1} + \mathop{\mathbf{K}}_{n=2}^{\infty} \frac{\gamma_n x^2}{1}, \tag{3.107a}$$

where

$$\gamma_{2m} = (2m-1)^2 k'^2 \quad \textit{and} \quad \gamma_{2m+1} = -(2m)^2 k^2. \tag{3.107b}$$

$$\int_0^\infty \operatorname{nd} u \, e^{-u/x} du = \frac{x}{1} + \mathop{\mathbf{K}}_{n=2}^{\infty} \frac{\gamma_n x^2}{1}, \tag{3.108a}$$

where

$$\gamma_{2m} = -(2m-1)^2 k^2 \quad \textit{and} \quad \gamma_{2m+1} = (2m)^2 k'^2. \tag{3.108b}$$

Next, take the formal Laplace transform in Definition 3.9 of the indicated Maclaurin series expansions from Definition 2.3. Let $m = 1, 2, 3, \ldots$. We then have the regular C-fraction expansions:

$$\int_0^\infty \operatorname{dc} u \, e^{-u/x} du = \frac{x}{1} + \mathop{\mathbf{K}}_{n=2}^{\infty} \frac{\gamma_n x^2}{1}, \tag{3.109a}$$

where

$$\gamma_{2m} = -(2m-1)^2 k'^2 \quad \textit{and} \quad \gamma_{2m+1} = -(2m)^2. \tag{3.109b}$$

$$\int_0^\infty \operatorname{nc} u \, e^{-u/x} du = \frac{x}{1} + \mathop{\mathbf{K}}_{n=2}^{\infty} \frac{\gamma_n x^2}{1}, \tag{3.110a}$$

where

$$\gamma_{2m} = -(2m-1)^2 \quad \textit{and} \quad \gamma_{2m+1} = -(2m)^2 k'^2. \tag{3.110b}$$

Proof: To obtain (3.105)–(3.108), we take (3.6), (3.7), (3.11), (3.12), and multiply both sides by x, set $x^2 = w$, and then identify by inspection what the γ_n's are in (3.103) and (3.104). We then replace (3.104) by (3.103), divide by x, and simplify. The expansions in (3.109) and (3.110) follow in the same way from (3.81) and (3.82), once we note that the left hand sides are formal Laplace transforms.

The regular C-fraction expansions in (3.107)–(3.110) are also a direct consequence of (3.105), (3.106), Lemma 3.8, and the modular transformations in (3.52b), (3.52c), (3.95b), and (3.95c), respectively. Apply Lemma 3.8 to the (3.91) case of the relations in (3.94) corresponding to these four cases of (3.90), while keeping in mind the regular C-fraction expansions in (3.105) and (3.106). □

The regular C-fraction expansions (3.105) and (3.106) appear in Rogers [207, p. 77], and Ramanujan was the first to obtain the regular C-fraction expansion equivalent to (3.107). Berndt [21, p. 165] obtains (3.107) by applying the appropriate modular transformations to (3.105). The rest of the regular C-fraction expansions in Theorem 3.11 appear to be new.

The associated continued fraction expansions in (3.8), (3.9), (3.13), (3.14), (3.84), and (3.85) can also be obtained directly from Theorem 3.11 in a simple manner. This alternate proof of (3.8) from (3.106) was also first given by Ismail and Masson in [111]. Their method of proof extends to the next three expansions in (3.9), (3.13), (3.14). The last two derivations of (3.84) and (3.85) from (3.116) and (3.117), respectively, are formal.

Keeping in mind the differentiation formulas for dn, cn, nd, cd, nc, dc it turns out that (3.8), (3.9), (3.13), (3.14), (3.84), (3.85) are a consequence of one integration by parts, Theorem 3.11, and the following identity for continued fractions that can be derived from Lemmas I and II in Rogers [207].

$$1 - \frac{1}{1+} \mathop{\mathbf{K}}_{n=2}^{\infty} \frac{c_n x^2}{1} == \frac{c_2 x^2}{1+(c_2+c_3)x^2+} \mathop{\mathbf{K}}_{n=2}^{\infty} \frac{-c_{2n-1}c_{2n}x^4}{1+(c_{2n}+c_{2n+1})x^2}. \tag{3.111}$$

Just apply Theorem 3.11 and Eq. (3.111) to the right-hand-sides of the following integration by parts identities.

$$\int_0^\infty \mathrm{sn}(u,k)\,\mathrm{cn}(u,k)\,e^{-u/x}\,du = \frac{1}{k^2}\left[1 - \frac{1}{x}\int_0^\infty \mathrm{dn}(u,k)\,e^{-u/x}\,du\right] \tag{3.112}$$

$$\int_0^\infty \mathrm{sn}(u,k)\,\mathrm{dn}(u,k)\,e^{-u/x}\,du = \left[1 - \frac{1}{x}\int_0^\infty \mathrm{cn}(u,k)\,e^{-u/x}\,du\right] \tag{3.113}$$

$$\int_0^\infty \frac{\mathrm{sn}(u,k)\,\mathrm{cn}(u,k)}{\mathrm{dn}^2(u,k)}\,e^{-u/x}\,du = \frac{-1}{k^2}\left[1 - \frac{1}{x}\int_0^\infty \mathrm{nd}(u,k)\,e^{-u/x}\,du\right] \tag{3.114}$$

$$\int_0^\infty \frac{\mathrm{sn}(u,k)}{\mathrm{dn}^2(u,k)}\, e^{-u/x}\, du = \frac{1}{k'^2}\left[1 - \frac{1}{x}\int_0^\infty \mathrm{cd}(u,k)\, e^{-u/x}\, du\right] \tag{3.115}$$

$$\int_0^\infty \frac{\mathrm{sn}(u,k)\,\mathrm{dn}(u,k)}{\mathrm{cn}^2(u,k)}\, e^{-u/x}\, du = -\left[1 - \frac{1}{x}\int_0^\infty \mathrm{nc}(u,k)\, e^{-u/x}\, du\right] \tag{3.116}$$

$$\int_0^\infty \frac{\mathrm{sn}(u,k)}{\mathrm{cn}^2(u,k)}\, e^{-u/x}\, du = \frac{-1}{k'^2}\left[1 - \frac{1}{x}\int_0^\infty \mathrm{dc}(u,k)\, e^{-u/x}\, du\right] \tag{3.117}$$

Theorems 3.2 and 3.10 have interesting special cases when $k=0$ and $k=1$. It is well-known [135, pp. 26 and 39] that

$$\mathrm{sn}(u,0) = \sin u, \quad \mathrm{cn}(u,0) = \cos u, \quad \mathrm{dn}(u,0) = 1, \tag{3.118}$$

$$\mathrm{sn}(u,1) = \tanh u, \quad \mathrm{cn}(u,1) = \mathrm{dn}(u,1) = \mathrm{sech}\, u. \tag{3.119}$$

That is, the Jacobi elliptic functions interpolate between circular and hyperbolic functions.

Equations (3.11) and (3.12) interpolate between the Laplace transform of $\cos u$, 1, and $\cosh u$ when $k = 0$ and $k = 1$. Equations (3.10) and (3.16) interpolate between the Laplace transform of $\sin u$ and $\sinh u$, and $\sin^2 u$ and $\sinh^2 u$, respectively, when $k = 0$ and $k = 1$.

The $k = 1$ case of (3.5) and (3.86) both give the Laplace transform of $\tanh u$. Furthermore, (3.86) interpolates between the formal Laplace transform of $\tan u$ and the Laplace transform of $\tanh u$ when $k = 0$ and $k = 1$. That is, we have

$$\int_0^\infty \tan u\, e^{-u/x}\, du = \frac{x^2}{1-2x^2+} \mathop{\mathbf{K}}_{n=2}^{\infty} \frac{-(2n-1)(2n-2)^2(2n-3)x^4}{1-2(2n-1)^2x^2}, \tag{3.120}$$

$$\int_0^\infty \tanh u\, e^{-u/x}\, du = \frac{x^2}{1+2x^2+} \mathop{\mathbf{K}}_{n=2}^{\infty} \frac{-(2n-1)(2n-2)^2(2n-3)x^4}{1+2(2n-1)^2x^2}. \tag{3.121}$$

When viewed as associated continued fraction expansions of Laplace transforms, the convergence conditions that would be required in Theorem 3.10 depend upon the methods used to interpret the integrals and/or associated continued fractions. First, Laplace transforms of a periodic function can be rewritten as an infinite sum of integrals over a fixed minimal period. In this case we only have to deal with at most 2 singularities of the integrand. Next, the Laplace transform integrals can be discussed in the context of the Hadamard integral in [9, Section 5, pp. 45–46]. Finally, we can consider multisection of the associated continued fractions. For example, look at the numerators and/or denominators of the even and/or odd partial quotients. For additional analytic work relating to associated continued fractions and Jacobi elliptic functions, see [9, 19, 21–23, 38, 44, 46, 47, 49, 57, 59, 60, 83, 109, 110, 112–116, 148, 218–220, 224, 231–235], [185, p. 330]. We do not pursue these matters further here.

4. Hankel and χ determinant evaluations

In this section we utilize Theorems 7.14 and 7.2 of [119, pp. 244–246; pp. 223–224], row and column operations, and modular transformations to deduce our Hankel and χ determinant evaluations from the continued fraction expansions of Section 3.

We start with the Hankel determinant evaluations in the following theorem.

Theorem 4.1. *Take $H_n^{(1)}(\{c_\nu\})$ and $H_n^{(2)}(\{c_\nu\})$ to be the $n \times n$ Hankel determinants in* (3.55). *Let the elliptic function polynomials* $(\text{elliptic})_m(k^2)$ *of k^2, with k the modulus, be determined by the coefficients of the Maclaurin series expansions in Definition* 2.3. *Let $n = 1, 2, 3, \ldots$. Then,*

$$H_n^{(1)}(\{(sn)_\nu(k^2)\}) = (k^2)^{\binom{n}{2}} \prod_{r=1}^{2n-1} r!, \tag{4.1}$$

$$H_n^{(1)}(\{(cn)_{\nu-1}(k^2)\}) = H_n^{(1)}(\{(dn)_{\nu-1}(k^2)\}) = (k^2)^{\binom{n}{2}} \prod_{r=1}^{n-1} (2r)!^2, \tag{4.2}$$

$$H_n^{(1)}(\{(sc)_\nu(k^2)\}) = H_n^{(1)}(\{(sd)_\nu(k^2)\}) = (k^2)^{\binom{n}{2}} \prod_{r=1}^{n} (2r-1)!^2, \tag{4.3}$$

$$H_n^{(1)}(\{(s/d)_\nu(k^2)\}) = (-1)^{\binom{n}{2}} [k^2(1-k^2)]^{\binom{n}{2}} \prod_{r=1}^{2n-1} r!, \tag{4.4}$$

$$\begin{aligned} H_n^{(1)}(\{(c/d)_{\nu-1}(k^2)\}) &= H_n^{(1)}(\{(nd)_{\nu-1}(k^2)\}) \\ &= (-1)^{\binom{n}{2}} [k^2(1-k^2)]^{\binom{n}{2}} \prod_{r=1}^{n-1} (2r)!^2, \end{aligned} \tag{4.5}$$

$$\begin{aligned} H_n^{(1)}(\{(sc/d^2)_\nu(k^2)\}) &= H_n^{(1)}(\{(s/d^2)_\nu(k^2)\}) \\ &= (-1)^{\binom{n}{2}} [k^2(1-k^2)]^{\binom{n}{2}} \prod_{r=1}^{n} (2r-1)!^2, \end{aligned} \tag{4.6}$$

$$H_n^{(1)}(\{(sn^2)_\nu(k^2)\}) = (k^2)^{\binom{n}{2}} \prod_{r=1}^{2n} r!, \tag{4.7}$$

$$H_n^{(1)}(\{(s^2/d^2)_\nu(k^2)\}) = (-1)^{\binom{n}{2}} [k^2(1-k^2)]^{\binom{n}{2}} \prod_{r=1}^{2n} r!, \tag{4.8}$$

$$H_n^{(1)}(\{(sc/d)_\nu(k^2)\}) = (k^2)^{2\binom{n}{2}} \prod_{r=1}^{2n-1} r!, \tag{4.9}$$

$$H_n^{(1)}(\{(s^2c^2/d^2)_\nu(k^2)\}) = (k^2)^{2\binom{n}{2}} \prod_{r=1}^{2n} r!, \tag{4.10}$$

$$H_n^{(1)}(\{(s/c)_\nu(k^2)\}) = (1-k^2)^{\binom{n}{2}} \prod_{r=1}^{2n-1} r!, \tag{4.11}$$

$$H_n^{(1)}(\{(d/c)_{\nu-1}(k^2)\}) = H_n^{(1)}(\{(nc)_{\nu-1}(k^2)\}) = (1-k^2)^{\binom{n}{2}} \prod_{r=1}^{n-1} (2r)!^2, \tag{4.12}$$

$$H_n^{(1)}(\{(sd/c^2)_\nu(k^2)\}) = H_n^{(1)}(\{(s/c^2)_\nu(k^2)\}) = (1-k^2)^{\binom{n}{2}} \prod_{r=1}^{n} (2r-1)!^2, \tag{4.13}$$

$$H_n^{(1)}(\{(s^2/c^2)_\nu(k^2)\}) = (1-k^2)^{\binom{n}{2}} \prod_{r=1}^{2n} r!, \tag{4.14}$$

$$H_n^{(1)}(\{(sd/c)_\nu(k^2)\}) = \prod_{r=1}^{2n-1} r!, \tag{4.15}$$

$$H_n^{(1)}(\{(s/cd)_\nu(k^2)\}) = (1-k^2)^{2\binom{n}{2}} \prod_{r=1}^{2n-1} r!, \tag{4.16}$$

$$H_n^{(1)}(\{(s^2d^2/c^2)_\nu(k^2)\}) = \prod_{r=1}^{2n} r!, \tag{4.17}$$

$$H_n^{(1)}(\{(s^2/c^2d^2)_\nu(k^2)\}) = (1-k^2)^{2\binom{n}{2}} \prod_{r=1}^{2n} r!, \tag{4.18}$$

$$H_n^{(2)}(\{(cn)_{\nu-1}(k^2)\}) = H_n^{(1)}(\{(cn)_\nu(k^2)\}) = (-1)^n (k^2)^{\binom{n}{2}} \prod_{r=1}^{n} (2r-1)!^2, \tag{4.19}$$

$$H_n^{(2)}(\{(dn)_{\nu-1}(k^2)\}) = H_n^{(1)}(\{(dn)_\nu(k^2)\}) = (-1)^n (k^2)^{\binom{n+1}{2}} \prod_{r=1}^{n} (2r-1)!^2, \tag{4.20}$$

$$\begin{aligned} H_n^{(2)}(\{(c/d)_{\nu-1}(k^2)\}) &= H_n^{(1)}(\{(c/d)_\nu(k^2)\}) \\ &= (-1)^{\binom{n+1}{2}} (k^2)^{\binom{n}{2}} (1-k^2)^{\binom{n+1}{2}} \prod_{r=1}^{n} (2r-1)!^2, \end{aligned} \tag{4.21}$$

$$\begin{aligned} H_n^{(2)}(\{(nd)_{\nu-1}(k^2)\}) &= H_n^{(1)}(\{(nd)_\nu(k^2)\}) \\ &= (-1)^{\binom{n}{2}} (k^2)^{\binom{n+1}{2}} (1-k^2)^{\binom{n}{2}} \prod_{r=1}^{n} (2r-1)!^2, \end{aligned} \tag{4.22}$$

$$H_n^{(2)}(\{(d/c)_{\nu-1}(k^2)\}) = H_n^{(1)}(\{(d/c)_\nu(k^2)\}) = (1-k^2)^{\binom{n+1}{2}} \prod_{r=1}^{n} (2r-1)!^2, \tag{4.23}$$

$$H_n^{(2)}(\{(nc)_{\nu-1}(k^2)\}) = H_n^{(1)}(\{(nc)_\nu(k^2)\}) = (1-k^2)^{\binom{n}{2}} \prod_{r=1}^{n} (2r-1)!^2, \tag{4.24}$$

Proof: Substitute the Maclaurin series in Definition 2.3 into the left hand sides of (3.5)–(3.18), and (3.80)–(3.89). Take the formal Laplace transform, multiply both sides by x, 1, or x^{-1}, where the Maclaurin series is of the form (3.91), (3.92), or (3.93), respectively, and then add 1 to both sides. The Hankel determinant evaluations in (4.1)–(4.18) are now immediate consequences of Eq. (3.60) of Theorem 3.4.

An alternate approach is to first establish (4.1), (4.2), (4.3), and (4.7) as above, and then deduce the rest of (4.1)–(4.18) from these by appealing to suitable modular transformations and Eqs. (3.66) and (3.67) of Lemma 3.6.

We start with a modular transformation such as (3.90) where A and B are nonzero constants, k_1 is a function of the modulus k, $g(u,k)$ is one of the Jacobi elliptic functions $\mathrm{sn}(u,k)$, $\mathrm{cn}(u,k)$, $\mathrm{dn}(u,k)$, $\mathrm{sn}(u,k)\,\mathrm{cn}(u,k)$, $\mathrm{sn}(u,k)\,\mathrm{dn}(u,k)$, $\mathrm{sn}^2(u,k)$, and $f(u,k)$ is one of the other quotients of Jacobi elliptic functions in the integrands of the (formal) Laplace transforms in (3.80)–(3.89), and (3.10)–(3.14), (3.16)–(3.18). Let the Maclaurin series expansions of $f(u,k)$ and $g(u,k)$ be given by one of (3.91), (3.92), and (3.93). Equating coefficients of powers of u in the Maclaurin series expansion of both sides of (3.90) gives

$$\bar{f}_m(k^2) = CB^{2m}\bar{g}_m(k_1^2), \tag{4.25}$$

where $m = 1, 2, 3, \ldots$, and either $\bar{f}_m = f_{m-1}$, $\bar{g}_m = g_{m-1}$, $C = AB^{-2}$ in (3.91); $\bar{f}_m = f_m$, $\bar{g}_m = g_m$, $C = AB^{-1}$ in (3.92); and $\bar{f}_m = f_m$, $\bar{g}_m = g_m$, $C = A$ in (3.93). The rest of our formulas in (4.1)–(4.18) are a direct consequence of

$$H_n^{(1)}(\{\bar{f}_\nu(k^2)\}) = C^n H_n^{(1)}(\{B^{2\nu}\bar{g}_\nu(k_1^2)\}), \tag{4.26}$$

Eqs. (3.66) and (3.67) of Lemma 3.6, and Eqs. (4.1)–(4.3), and (4.7).

The necessary cases of (3.90) are given by (3.50), (3.52a)–(3.52c), (3.95a)–(3.95c), (3.96), (3.97), (3.98), (3.99), (3.100), squaring both sides of (3.50) and (3.52a), composing each of (3.101) and (3.102) with (3.50) to give

$$\frac{\mathrm{sn}(u,k)\,\mathrm{dn}(u,k)}{\mathrm{cn}(u,k)} = -(k+ik')\mathrm{sn}((ik'-k)u, 1-2k(k+ik')), \tag{4.27}$$

and

$$\frac{\mathrm{sn}^2(u,k)\,\mathrm{dn}^2(u,k)}{\mathrm{cn}^2(u,k)} = -[1-2k(k+ik')]\mathrm{sn}^2((ik'-k)u, 1-2k(k+ik')), \tag{4.28}$$

and finally, the combinations of (3.52a), (3.52b), (3.52c) given by

$$\frac{\mathrm{sn}(u,k)\,\mathrm{cn}(u,k)}{\mathrm{dn}^2(u,k)} = \frac{1}{k'}\,\mathrm{sn}(k'u, ik/k')\,\mathrm{cn}(k'u, ik/k'), \tag{4.29}$$

$$\frac{\mathrm{sn}(u,k)}{\mathrm{dn}^2(u,k)} = \frac{1}{k'}\,\mathrm{sn}(k'u, ik/k')\,\mathrm{dn}(k'u, ik/k'). \tag{4.30}$$

To establish (4.19)–(4.24) from Theorem 3.11, we recall the relevant Maclaurin series in (2.62)–(2.66), apply (3.79) to the left hand side of each regular C-fraction expansion in Theorem 3.11, multiply by x, set $x^2 = w$, then add 1 to both sides. We obtain

$$1 + \sum_{m=1}^{\infty}(elliptic)_{m-1}(k^2)w^m = 1 + \frac{w}{1} + \mathop{\mathbf{K}}_{n=2}^{\infty}\frac{\gamma_n w}{1}, \tag{4.31}$$

where $\{\gamma_n\}$ is determined by Theorem 3.11. Equations (4.19)–(4.24) now follow immediately from (3.65) of Theorem 3.5 and the $H_n^{(1)}(\{c_{\nu-1}\})$ evaluations in (4.2), (4.5), and (4.12). □

The Hankel determinant evaluations in (4.1), (4.2), (4.7), (4.19), and (4.20) were also obtained earlier from (3.60) in [2, Eqs. (10.13), (10.17), (10.14), (10.18), pp. 97–99]. The rest appear to be new.

Note that the product sides of (4.15) and (4.17) are independent of k. Analogous independence results for classical orthogonal polynomials appear in [80, 81]. For a more elementary type of independence result of Sylvester for Hankel determinants, see [177, Vol. III, pp. 316–317].

The $H_n^{(1)}$ evaluations in (4.19)–(4.24) are not utilized directly in the proof of the χ_n determinant evaluations in Theorem 4.2. However, they are equivalent to the Hankel determinant evaluations in (4.3), (4.6), and (4.13) which are used in the proof of Theorem 4.2. To see this equivalence first equate coefficients of u^N in both sides of the Maclaurin series expansion of the differentiation formulas for dn, cn, nd, cd, nc, dc. The evaluations in (4.3), (4.6), and (4.13) then follow immediately from factoring suitable constants from the resulting determinants and appealing to (4.19)–(4.24). Note that the differentiation formula for nc thus explains why the evaluations in (4.13) and (4.24) have the same answer.

It is not hard to see from Theorems 3.2, 3.10, 4.1, and Eq. (3.61) of Theorem 3.4 that we have the following theorem.

Theorem 4.2. *Take $\chi_n(\{c_\nu\})$ to be the $n \times n$ determinant in* (3.56). *Let the elliptic function polynomials* (elliptic)$_m(k^2)$ *of k^2, with k the modulus, be determined by the coefficients of the Maclaurin series expansions in Definition* 2.3. *Let $n = 1, 2, 3, \ldots$. Then,*

$$\chi_n(\{(sn)_\nu(k^2)\}) = -\tfrac{n(4n^2-1)}{3}(k^2)^{\binom{n}{2}}(1+k^2)\prod_{r=1}^{2n-1} r!, \tag{4.32}$$

$$\chi_n(\{(cn)_{\nu-1}(k^2)\}) = -\tfrac{n(2n-1)}{3}(k^2)^{\binom{n}{2}}[2n(1+k^2)+(1-2k^2)]\prod_{r=1}^{n-1}(2r)!^2, \tag{4.33}$$

$$\chi_n(\{(dn)_{\nu-1}(k^2)\}) = -\tfrac{n(2n-1)}{3}(k^2)^{\binom{n}{2}}[2n(1+k^2)-(2-k^2)]\prod_{r=1}^{n-1}(2r)!^2, \tag{4.34}$$

$$\chi_n(\{(sc)_\nu(k^2)\}) = -\tfrac{n(2n+1)}{3}(k^2)^{\binom{n}{2}}[2n(1+k^2)+(2-k^2)]\prod_{r=1}^{n}(2r-1)!^2, \tag{4.35}$$

$$\chi_n(\{(sd)_\nu(k^2)\}) = -\tfrac{n(2n+1)}{3}(k^2)^{\binom{n}{2}}[2n(1+k^2)-(1-2k^2)]\prod_{r=1}^{n}(2r-1)!^2, \tag{4.36}$$

$$\chi_n(\{(s/d)_\nu(k^2)\}) = -(-1)^{\binom{n}{2}}\cdot\tfrac{n(4n^2-1)}{3}[k^2(1-k^2)]^{\binom{n}{2}}(1-2k^2)\prod_{r=1}^{2n-1} r!, \tag{4.37}$$

$$\begin{aligned}\chi_n(\{(c/d)_{\nu-1}(k^2)\}) &= -(-1)^{\binom{n}{2}}\cdot\tfrac{n(2n-1)}{3}[k^2(1-k^2)]^{\binom{n}{2}}\\ &\quad\times[2n(1-2k^2)+(1+k^2)]\prod_{r=1}^{n-1}(2r)!^2,\end{aligned} \tag{4.38}$$

$$\chi_n(\{(nd)_{\nu-1}(k^2)\}) = -(-1)^{\binom{n}{2}} \cdot \tfrac{n(2n-1)}{3}[k^2(1-k^2)]^{\binom{n}{2}} \times [2n(1-2k^2)-(2-k^2)]\prod_{r=1}^{n-1}(2r)!^2, \tag{4.39}$$

$$\chi_n(\{(sc/d^2)_{\nu}(k^2)\}) = -(-1)^{\binom{n}{2}} \cdot \tfrac{n(2n+1)}{3}[k^2(1-k^2)]^{\binom{n}{2}} \times [2n(1-2k^2)+(2-k^2)]\prod_{r=1}^{n}(2r-1)!^2, \tag{4.40}$$

$$\chi_n(\{(s/d^2)_{\nu}(k^2)\}) = -(-1)^{\binom{n}{2}} \cdot \tfrac{n(2n+1)}{3}[k^2(1-k^2)]^{\binom{n}{2}} \times [2n(1-2k^2)-(1+k^2)]\prod_{r=1}^{n}(2r-1)!^2, \tag{4.41}$$

$$\chi_n(\{(sn^2)_{\nu}(k^2)\}) = -\tfrac{2n(n+1)(2n+1)}{3}(k^2)^{\binom{n}{2}}(1+k^2)\prod_{r=1}^{2n} r!, \tag{4.42}$$

$$\chi_n(\{(s^2/d^2)_{\nu}(k^2)\}) = -(-1)^{\binom{n}{2}} \cdot \tfrac{2n(n+1)(2n+1)}{3}[k^2(1-k^2)]^{\binom{n}{2}}(1-2k^2)\prod_{r=1}^{2n} r!, \tag{4.43}$$

$$\chi_n(\{(sc/d)_{\nu}(k^2)\}) = -\tfrac{2n(4n^2-1)}{3}(k^2)^{2\binom{n}{2}}(2-k^2)\prod_{r=1}^{2n-1} r!, \tag{4.44}$$

$$\chi_n(\{(s^2c^2/d^2)_{\nu}(k^2)\}) = -\tfrac{4n(n+1)(2n+1)}{3}(k^2)^{2\binom{n}{2}}(2-k^2)\prod_{r=1}^{2n} r!, \tag{4.45}$$

$$\chi_n(\{(s/c)_{\nu}(k^2)\}) = \tfrac{n(4n^2-1)}{3}(1-k^2)^{\binom{n}{2}}(2-k^2)\prod_{r=1}^{2n-1} r!, \tag{4.46}$$

$$\chi_n(\{(d/c)_{\nu-1}(k^2)\}) = \tfrac{n(2n-1)}{3}(1-k^2)^{\binom{n}{2}}[2n(2-k^2)-(1+k^2)]\prod_{r=1}^{n-1}(2r)!^2, \tag{4.47}$$

$$\chi_n(\{(nc)_{\nu-1}(k^2)\}) = \tfrac{n(2n-1)}{3}(1-k^2)^{\binom{n}{2}}[2n(2-k^2)-(1-2k^2)]\prod_{r=1}^{n-1}(2r)!^2, \tag{4.48}$$

$$\chi_n(\{(sd/c^2)_{\nu}(k^2)\}) = \tfrac{n(2n+1)}{3}(1-k^2)^{\binom{n}{2}}[2n(2-k^2)+(1-2k^2)]\prod_{r=1}^{n}(2r-1)!^2, \tag{4.49}$$

$$\chi_n(\{(s/c^2)_{\nu}(k^2)\}) = \tfrac{n(2n+1)}{3}(1-k^2)^{\binom{n}{2}}[2n(2-k^2)+(1+k^2)]\prod_{r=1}^{n}(2r-1)!^2, \tag{4.50}$$

$$\chi_n(\{(s^2/c^2)_{\nu}(k^2)\}) = \tfrac{2n(n+1)(2n+1)}{3}(1-k^2)^{\binom{n}{2}}(2-k^2)\prod_{r=1}^{2n} r!, \tag{4.51}$$

$$\chi_n(\{(sd/c)_{\nu}(k^2)\}) = \tfrac{2n(4n^2-1)}{3}(1-2k^2)\prod_{r=1}^{2n-1} r!, \tag{4.52}$$

$$\chi_n(\{(s/cd)_\nu(k^2)\}) = \tfrac{2n(4n^2-1)}{3}(1-k^2)^{2\binom{n}{2}}(1+k^2)\prod_{r=1}^{2n-1} r!, \tag{4.53}$$

$$\chi_n(\{(s^2d^2/c^2)_\nu(k^2)\}) = \tfrac{4n(n+1)(2n+1)}{3}(1-2k^2)\prod_{r=1}^{2n} r!, \tag{4.54}$$

$$\chi_n(\{(s^2/c^2d^2)_\nu(k^2)\}) = \tfrac{4n(n+1)(2n+1)}{3}(1-k^2)^{2\binom{n}{2}}(1+k^2)\prod_{r=1}^{2n} r!. \tag{4.55}$$

The same Maclaurin series expansions of modular transformations that were utilized in the second proof of Theorem 4.1 can also be combined with equations (3.68) and (3.69) of Lemma 3.6 to deduce the rest of Eqs. (4.32)–(4.55) from (4.32)–(4.36), and (4.42).

All of the nonconstant factors in the products on the right hand sides of the identities in Theorem 4.2, which are not powers of k^2 or $(1-k^2)$, are linear combinations of certain subsets of $(1+k^2)$, $(2-k^2)$, and $(1-2k^2)$. The 24 identities in Theorem 4.2 also have a number of elegant symmetries involving these last three expressions. In Section 5, the $(1+k^2)$, $(2-k^2)$, and $(1-2k^2)$ correspond to Lambert series.

Keeping in mind (3.118) and (3.119), we see that Theorems 4.1 and 4.2 have interesting special cases when $k=0$ and $k=1$. In this paper we need the $k=0$ cases in the following theorem.

Theorem 4.3. *Take $H_n^{(1)}(\{c_\nu\})$, $H_n^{(2)}(\{c_\nu\})$, and $\chi_n(\{c_\nu\})$ to be the $n\times n$ determinants in Definition 3.3. Let the Bernoulli numbers B_n and Euler numbers E_n be defined by (2.60) and (2.61), respectively. Let $n=1,2,3,\ldots$. Then,*

$$H_n^{(1)}\left(\left\{(-1)^{\nu-1}\frac{(2^{2\nu}-1)}{4\nu}\cdot|B_{2\nu}|\right\}\right) = 2^{-(2n^2+n)}\prod_{r=1}^{2n-1} r!, \tag{4.56}$$

$$H_n^{(1)}\left(\left\{(-1)^{\nu}\frac{(2^{2\nu+2}-1)}{4(\nu+1)}\cdot|B_{2\nu+2}|\right\}\right) = (-1)^n 2^{-(2n^2+3n)}\prod_{r=1}^{2n} r!, \tag{4.57}$$

$$H_n^{(1)}(\{(-1)^{\nu-1}\cdot\tfrac{1}{4}\cdot|E_{2\nu-2}|\}) = 2^{-2n}\prod_{r=1}^{n-1}(2r)!^2, \tag{4.58}$$

$$\begin{aligned} H_n^{(2)}(\{(-1)^{\nu-1}\cdot\tfrac{1}{4}\cdot|E_{2\nu-2}|\}) &= H_n^{(1)}(\{(-1)^{\nu}\cdot\tfrac{1}{4}\cdot|E_{2\nu}|\}) \\ &= (-1)^n 2^{-2n}\prod_{r=1}^{n}(2r-1)!^2, \end{aligned} \tag{4.59}$$

$$\chi_n\left(\left\{(-1)^{\nu-1}\frac{(2^{2\nu}-1)}{4\nu}\cdot|B_{2\nu}|\right\}\right) = -\tfrac{n(4n^2-1)}{3}2^{-(2n^2+n+1)}\prod_{r=1}^{2n-1} r!, \tag{4.60}$$

$$\chi_n\left(\left\{(-1)^{\nu}\frac{(2^{2\nu+2}-1)}{4(\nu+1)}\cdot|B_{2\nu+2}|\right\}\right) = (-1)^{n-1}\cdot\tfrac{n(n+1)(2n+1)}{3}2^{-(2n^2+3n)}\prod_{r=1}^{2n} r!, \tag{4.61}$$

$$\chi_n(\{(-1)^{\nu-1}\cdot\tfrac{1}{4}\cdot|E_{2\nu-2}|\}) = -\tfrac{n(2n-1)(4n-1)}{3}2^{-2n}\prod_{r=1}^{n-1}(2r)!^2, \tag{4.62}$$

$$\chi_n(\{(-1)^{\nu}\cdot\tfrac{1}{4}\cdot|E_{2\nu}|\}) = -(-1)^n\tfrac{n(2n+1)(4n+1)}{3}2^{-2n}\prod_{r=1}^{n}(2r-1)!^2. \tag{4.63}$$

Proof: Since $\tan u = \mathrm{sc}(u,0)$, we find after equating powers of u in the $u \mapsto uz$ case of (2.56), and the $k=0$ case of (2.65) that

$$(-1)^{m-1}\frac{(2^{2m}-1)}{4m}\cdot|B_{2m}| = \frac{(-1)^{m-1}}{2^{2m+1}}\cdot(s/c)_m(0), \tag{4.64}$$

for $m = 1, 2, 3, \ldots$. Equation (4.56) now follows by (3.66) and the $k=0$ case of (4.11).

Next, from $\sec^2 u = 1 + \mathrm{sc}^2(u,0)$, we obtain after equating powers of u in the $u \mapsto uz$ case of (2.57), and the $k=0$ case of (2.67) that

$$(-1)^m\frac{(2^{2m+2}-1)}{4(m+1)}\cdot|B_{2m+2}| = \frac{(-1)^m}{2^{2m+3}}\cdot(s^2/c^2)_m(0), \tag{4.65}$$

for $m = 1, 2, 3, \ldots$. Equation (4.57) now follows by (3.66) and the $k=0$ case of (4.14).

By $\sec u = \mathrm{nc}(u,0)$, we find after equating powers of u in the $u \mapsto uz$ case of (2.58), and the $k=0$ case of (2.66) that

$$(-1)^{m-1}\cdot\tfrac{1}{4}\cdot|E_{2m-2}| = (-1)^{m-1}\cdot\tfrac{1}{4}\cdot(nc)_{m-1}(0), \tag{4.66}$$

for $m = 1, 2, 3, \ldots$. Equation (4.58) now follows by (3.67) and the $k=0$ case of (4.12). We establish (4.59) the same way we obtained (4.58). Equation (4.59) is an immediate consequence of (4.66), (3.71), and the $k=0$ case of (4.24).

Equation (4.60) is immediate from (4.64), (3.68), and the $k=0$ case of (4.46). Equation (4.61) follows from (4.65), (3.68), and the $k=0$ case of (4.51). Equation (4.62) is immediate from (4.66), (3.69), and the $k=0$ case of (4.48).

By $\sec u \tan u = \mathrm{sc}(u,0)\,\mathrm{nc}(u,0)$, it follows after equating powers of u in the $u \mapsto uz$ case of (2.59), and the $k=0$ case of (2.78) that

$$(-1)^m\cdot\tfrac{1}{4}\cdot|E_{2m}| = (-1)^m\cdot\tfrac{1}{4}\cdot(s/c^2)_m(0), \tag{4.67}$$

for $m = 1, 2, 3, \ldots$. Equation (4.63) is now immediate from (4.67), (3.68), and the $k=0$ case of (4.50). □

In addition to our proofs of Theorems 4.1–4.3 we have used Mathematica [251] to directly verify a number of these determinant evaluations up to $n = 15$. We computed the various polynomials $(elliptic)_m(k^2)$ by utilizing Dumont's recursions and symmetries of the ordinary and symmetric Schett polynomials of x, y, and z from [58, Eq. (2.1), p. 4; Corollaire 2.2, p. 6; Eq. (4.1), p. 11; Proposition 4.1, p. 11]. The general ordinary Schett polynomials were first introduced by Schett [208, 209] in a slightly different form. (The fourth Schett polynomial X_4 appeared much earlier as $T_\alpha^{(\mathrm{IV})}(\nu)$ in [186, p. 84]). In 1926,

Mitra [170] employed classical methods to compute explicit Taylor series expansions of $\mathrm{sn}(u,k)$, and $\mathrm{cn}(u,k)$ and $\mathrm{dn}(u,k)$ about $u=0$, as far as u^{21} and u^{18}, respectively. Bell, in [15–18], utilized a more direct and more practicable method to explicitly compute these series and their reciprocals. Wrigge [253] obtained Maclaurin series expansions of $\mathrm{sn}(u,k)$ in terms of k^2. Dumont's bimodular functions in [58] are very similar to but slightly more symmetric than Abel's elliptic functions ϕ, f, and F introduced in [1]. Abel's functions are related to Dumont's functions as follows:

$$\phi(u;a,b)=\mathrm{sn}(u;ia,b),\quad f(u;a,b)=\mathrm{cn}(u;ia,b),\quad F(u;a,b)=\mathrm{dn}(u;ia,b). \tag{4.68}$$

The determinants $H_n^{(m)}(\{c_\nu\})$ in (3.55) are known in the literature as Hankel, Turánian, or persymmetric determinants, and sometimes as determinants of recurrent type. They are also called "Turánians" by Karlin and Szegő [122, p. 5]. An excellent survey of the classical literature on these determinants can be found in Muir's books and articles [177–181], and Krattenthaler's summary in [125, pp. 20–23; pp. 46–48]. For more details on [177], see [177, Vol. I, pp. 485–487], [177, Vol. II, pp. 324–357; pp. 461–462], [177, Vol. III, pp. 309–326; pp. 469–473], [177, Vol. IV, pp. 312–331; pp. 464–465]. For classical work on the related determinants known as recurrents, see [177, Vol. II, pp. 210–211], [177, Vol. III, pp. 208–247], [177, Vol. IV, pp. 224–241]. Sylvester [177, Vol. II, p. 341] was the first to use the term persymmetric. We prefer the more common term of Hankel determinant, in view of Hankel's very important work described in [177, Vol. III, pp. 312–316]. Additional and/or more recent research involving or directly related to Hankel determinants can be found in [2, 10, 13, 29, 30, 35, 36, 40, 44–47, 49, 50, 52–55, 62, 70, 72–75, 78, 80–82, 92, 97, 98, 105, 106, 109–116, 119, 122, 124, 125, 133, 135–138, 149, 153, 156, 157, 174–176, 181, 185, 190–193, 196–200, 206, 207, 210, 220, 221, 226, 236–241, 243, 245, 246, 248, 250, 255–258].

Heilermann's correspondences in Theorems 3.4 and 3.5, and similar results are well-known. For example, works that involved both associated continued fraction and regular C-fraction correspondences include the following: Rogers [207, pp. 72–74], Datta [54, pp. 109–110; p. 127], Perron [185, Satz 5, pp. 304–305; Satz 11, pp. 324–325], Jones and Thron [119, pp. 244–246; pp. 223–224], and D. and G. Chudnovsky [47, pp. 140–149]. For the associated continued fraction correspondence we have: Szász [221], Beckenbach et al. [13, p. 6], Viennot [245], Flajolet [70, p. 152], Goulden and Jackson [92, Example 5.2.21, p. 312; pp. 528–529], Hendriksen and Van Rossum [106, p. 321], and Zeng [257, p. 381]. Finally, for the regular C-fraction correspondence we note: Muir [174–176], Frobenius [72], Stieltjes [220], Wall [248, Eq. (44.7), p. 172; see also pp. 399–413], and Lorentzen and Waadeland [149, Eq. (2.5.2), p. 257].

Heilermann [103, 104] seems to be the first to discover and prove the explicit determinental formulas in Theorems 3.4 and 3.5 for expanding a formal power series into either an associated continued fraction or a regular C-fraction. However, special cases and/or a discussion of the general methods he used to arrive at his expansions (without his closed formulas) appeared earlier. For example, see page 7 of [119] and the survey in [31, pp. 190–193] for these matters, as well as references to some of the later work involving Theorems 3.4 and 3.5. The papers [174–176] of Muir, which include a rediscovery of many of Heilermann's results, also provides additional information on the earlier work of Euler, Gauß, and Stern. The paper [176] has many interesting explicit regular C-fraction expansions. Stern [216, pp. 245–259], [217] knew of these general methods and applied them to particular series. Gauß [119, pp. 198–214] obtained a simple, direct special case of [103, 104] by finding explicit regular C-fraction expansions for the ratio $F(a, b; c; z)/F(a, b+1; c+1; z)$ of hypergeometric functions, and of $F(a, 1; c; z)$. Euler's general method in [68, p. 138] leads to general T-fraction [119, Eq. (A.28), p. 390] expansions instead of regular C-fraction expansions. For additional related information on Euler's work on continued fractions, see [31, pp. 97–109], [65, 66, 68], and [67, chapter 18]. Further comments on Muir's papers [174, 176] appear in [31, pp. 181–182].

The most common use of Theorem 3.4 is to first obtain (3.57) and then appeal to (3.60) to compute the Hankel determinants $H_n^{(1)}(\{c_\nu\})$. Similar calculations are carried out in [2, 13, 70, 92, 106, 221, 245, 257]. On the other hand, continued fraction expansions analogous to (3.57) and/or (3.62) are derived in [47, 54] by first directly evaluating suitable Hankel determinants in order to obtain the required constants such as (3.59) and/or (3.64) in the continued fraction. Further evaluations of Hankel determinants that do not seem to follow from known results about continued fractions or orthogonal polynomials appear in [105, Property 14] and [75, Section 4]. The paper of Robbins [206] contains yet a different kind of Hankel determinant evaluation that is related to those of Cauchy [153, Example 6, p. 67], Frobenius and Stickelberger [73, 74], Muir [181, p. 352; Example 29 and 31, p. 357; Example 43, p. 360], and D. V. and G. V. Chudnovsky [47, pp. 143, 144, 147].

The method of first obtaining (3.57) and then appealing to (3.60) or (3.61), and/or deriving (3.62) and then using (3.65) is responsible for Theorems 4.1, 4.2, and (indirectly) Theorem 4.3. The Hankel determinant evaluations in (4.1), (4.2), (4.7), (4.19), and (4.20) were also obtained earlier from (3.60) in [2, pp. 97–99]. Equations (4.56) and (4.58) are equivalent to results which Flajolet described in [70, pp. 149, 152]. Flajolet refers to applying a formula similar to (3.60) to the associated continued fraction expansions for $\sec u$ and $\tan u$ in [70, Table 1, p. 149]. Flajolet's product formulas in [70, Table 1, p. 149] for Hankel determinants of Bell numbers and Hankel determinants of the number of derangements also appear in [55, 190, 192]. Flajolet also points out that there is a similar connection between his version of (3.60) and the classical regular C-fraction expansion equivalent to (3.105) in [69, p. 149]. However, (3.64) is not mentioned. Equations (4.58) and (4.59) are equivalent to Eqs. (3.52) and (3.53), respectively, in [125, Theorem 52, p. 46]. In addition, Eq. (4.58) is treated in [2, 138, 191–193]. Equations (4.58) and (4.59) are also equivalent to a (corrected) version,

$$\det(c'_{r+s})_{0\leq r,s\leq n-1} \equiv \det(|E_{r+s}|)_{0\leq r,s\leq n-1} = \prod_{r=0}^{n-1}(r!)^2, \tag{4.69}$$

of the second formula on page 323 of [157]. A correct version of the third formula on page 323 of [157] should be related to Eqs. (4.56) and (4.57). Very few $\chi_n(\{c_\nu\})$ determinants such as (3.56) have been explicitely evaluated in the literature for any $\{c_\nu\}$. One such example appears in [13, p. 6] where $c_\nu := P_{\nu-1}(x)$ is the $(\nu-1)$-st Legendre polynomial.

The determinant evaluations, other than (4.1), (4.2), (4.7), (4.19), (4.20), (4.56)–(4.59) in Theorems 4.1–4.3 do not appear to have been written down in the literature before.

One of the deepest and most interesting investigations in the literature of Hankel determinants is the work of Karlin and Szegő in [122]. In order to generalize Turán's [222, 225] inequality for Legendre polynomials they showed that the $2m$ by $2m$ Hankel determinants, whose entries are classical orthogonal polynomials (Legendre, ultraspherical, Laguerre, Hermite) of x, has a constant sign $(-1)^m$ for all x in a suitable interval. For $2m+1$ by $2m+1$ Hankel determinants of these polynomials they generalize the well known oscillation property of the orthogonal polynomials involved. A crucial part of their analysis was to express each of these ℓ by ℓ Hankel determinants as a constant multiple of $(x^2-1)^N$, x^N, or 1 times the n by n Wronskian determinant of certain orthogonal polynomials of another class (subject to a change of variables and normalization). This reduced their analysis to that of the sign of a Wronskian determinant, which is easier. These relations between Hankel determinants and Wronskians are given by [122, Eq. (12.1), p. 47; Eq. (14.1), p. 53; Eq. (16.1), pp. 60–61; Eq. (18.2), p. 69]. Karlin and Szegő describe in [122, pp. 4–5] how these relations are partially motivated by the duality between m and x of the classical orthogonal polynomials $\{Q_m(x)\}$. By setting $n=0$ in the above relations, Karlin and Szegő express certain of their ℓ by ℓ Hankel determinants as a constant multiple of $(x^2-1)^N$ or x^N, since a 0 by 0 Wronskian determinant equals 1. For example, see [122, Eq. (12.3), p. 47; Eq. (14.3), p. 53; Eq. (16.5), p. 61]. Karlin and Szegő's results from [122] are discussed in [156, Sections 9.1–9.2, pp. 133–135].

Karlin and Szegő's proofs in [122] of their relations between Hankel determinants and Wronskians do not clearly exhibit a general algebraic transformation, which when specialized in the context of suitable properties of the orthogonal polynomials under consideration, yields their results. Recently, Leclerc [137, 138] utilized Lascoux's [133] Schur function approach to orthogonal polynomials, the 'master identity' of Turnbull on minors of a matrix [226, p. 48], [135], and additional identities from [136] involving staircase Schur functions, to put together an extremely elegant derivation of a general algebraic transformation which specializes to give Karlin and Szegő's relations between Hankel determinants and Wronskians. Leclerc points out that this technique of realizing identities between orthogonal polynomials as specializations of identities for symmetric functions is motivated by Littlewood's [145, chapter 7] important idea that any sequence $\{a_m\}$, $m \geq 1$ of elements of a commutative ring R can be regarded as the sequence of complete homogeneous symmetric functions of a fictitious set of variables E: $S_m(E) = a_m$. Leclerc's general algebraic identity states that a Wronskian of orthogonal polynomials is proportional to a Hankel determinant whose elements form a new sequence of polynomials. In order to complete his derivation of Karlin and Szegő's relations, Leclerc verifies that in each case considered by Karlin and Szegő, this new sequence is transformed by a suitable change of variables and normalization into another class of classical orthogonal polynomials. Leclerc reduces these verifications to algebraic properties discovered by Burchnall [35] of the classical orthogonal polynomials.

It would be interesting to extend Karlin and Szegő's relations between determinants of classical orthogonal polynomials of Hankel and Wronskian type to a Jacobi elliptic function setting in which the elements of the Hankel determinants are functions of the Maclaurin series coefficients in Definition 2.3. This could provide more insights into the identities and number theory discussed in this paper. A hint in this direction is provided by such a relation in [122, Eq. (12.1), p. 47] involving Legendre polynomials and the complicated explicit formulas from [252, pp. 556–557] which express the Maclaurin series coefficients of $\operatorname{sn}(x,k)$ and $\operatorname{sn}^2(x,k)$ as sums of products of Legendre polynomials. One direct approach to this problem is to apply Theorem 1 of [138] to the orthogonal polynomials in [38, 115]. A second approach starts by seeking a continued fraction/lattice paths understanding of Theorem 1 of [138]. This combinatorial analysis should be directly related to [69, pp. 148–150; pp. 152–154], [70, 71, 133, 245, 246], and [82, p. 304]. In all the above analysis it is important to keep in mind the 'discrete Wronskian' in [122, Eq. (1.3), p. 5]. Roughly speaking, we want to use [137, 138] as a bridge between [122], this paper, and [69]. These connections deserve much further study.

In a different direction than [137, 138], it is instructive to equate the evaluations of the Hankel determinant of ultraspherical polynomials in Eq. (20) of [13, p. 5] with that in Eq. (14.3) of [122, p. 53]. This immediately gives an elegant evaluation of an integral equivalent to the classical case of Selberg's multiple beta integral [152, p. 992] in which the Vandermonde determinant in the integrand occurs to the second power. It would be interesting to see what other cases of Selberg's integral arise in a similar fashion from [122, 138].

5. The determinant form of sums of squares identities

In this section we combine Theorems 2.4, 4.1–4.3, some row and column operations from Lemma 3.6, and well-known relations between classical theta functions/Lambert series and the elliptic function parameters z and k, to derive first the single determinant, and then sum of determinants, form of our infinite families of sums of squares and related identities. We also provide (see Theorem 5.19 below) our generalization to infinite families all 21 of Jacobi's [117, Sections 40–42] explicitly stated degree 2, 4, 6, 8 Lambert series expansions of classical theta functions. An elegant generating function involving the number of ways of writing N as a sum of $2n$ squares and $(2n)^2$ triangular numbers appears in Corollary 5.15.

Throughout this section we need the Bernoulli numbers B_n and Euler numbers E_n, defined by (2.60) and (2.61), respectively. We also use the notation $I_n := \{1, 2, \ldots, n\}$, $\|S\|$ is the cardinality of the set S, and $\det(M)$ is the determinant of the $n \times n$ matrix M.

As background, we begin with two lemmas. The first is an inclusion-exclusion computation for Hankel determinants, and the second provides the necessary elliptic function parameter relations.

We first establish the inclusion-exclusion lemma.

Lemma 5.1. *Let $v_1, \ldots, v_{2n-1}$ and $w_1, \ldots, w_{2n-1}$ be indeterminate, and let $n = 1, 2, 3, \ldots$. Suppose that the (i, j) entries of the $n \times n$ matrices (w_{i+j-1}) and (v_{i+j-1}) are w_{i+j-1} and v_{i+j-1}, respectively, and that $M_{n,S}$ is the $n \times n$ matrix whose i-th row is*

$$v_i + w_i,\ v_{i+1} + w_{i+1}, \ldots, v_{i+n-1} + w_{i+n-1}, \quad \text{if } i \in S,$$

and

$$v_i,\ v_{i+1}, \ldots, v_{i+n-1}, \quad \text{if } i \notin S. \tag{5.1}$$

Then

$$\det(w_{i+j-1}) = \sum_{\emptyset\subseteq S\subseteq I_n} (-1)^{n-\|S\|} \det(M_{n,S}) \tag{5.2}$$

$$= (-1)^n \det(v_{i+j-1}) + \sum_{p=1}^{n} (-1)^{n-p} \sum_{\substack{\emptyset\subseteq S\subseteq I_n \\ \|S\|=p}} \det(M_{n,S}). \tag{5.3}$$

Proof: We start with the sum in (5.2) and expand $\det(M_{n,S})$ along each row corresponding to $i \in S$. For convenience, let $B_{n,T}$ denote the $n \times n$ matrix whose i-th row is

$$w_i, w_{i+1}, \ldots, w_{i+n-1}, \quad \text{if } i \in T \quad \text{and} \quad v_i, v_{i+1}, \ldots, v_{i+n-1}, \quad \text{if } i \notin T. \tag{5.4}$$

Expanding each row of $\det(M_{n,S})$ corresponding to $i \in S$ gives

$$\det(M_{n,S}) = \sum_{\emptyset\subseteq T\subseteq S} \det(B_{n,T}). \tag{5.5}$$

Substituting (5.5) into the sum in (5.2) and then interchanging summation yields

$$\sum_{\emptyset\subseteq S\subseteq I_n} (-1)^{n-\|S\|} \det(M_{n,S}) = \sum_{\emptyset\subseteq T\subseteq I_n} \det(B_{n,T}) \sum_{T\subseteq S\subseteq I_n} (-1)^{n-\|S\|}. \tag{5.6}$$

We use the binomial theorem to simplify the inner sum in the right-hand side of (5.6). Let $\|T\| = p$, for $p = 0, 1, 2, \ldots, n$. We then have

$$\begin{aligned}\sum_{T\subseteq S\subseteq I_n} (-1)^{n-\|S\|} &= \sum_{\nu=p}^{n} (-1)^{n-\nu} \binom{n-p}{\nu-p} \\ &= (-1)^{n-p} \sum_{\nu=0}^{n-p} (-1)^{\nu} \binom{n-p}{\nu} = (-1)^{n-p}(1-1)^{n-p} \\ &= 0, \quad \text{if } n \neq p, \quad \text{and} \quad 1, \quad \text{if } n = p.\end{aligned} \tag{5.7}$$

It now follows that (5.6) becomes

$$\begin{aligned}\sum_{\emptyset\subseteq S\subseteq I_n} (-1)^{n-\|S\|} \det(M_{n,S}) &= \sum_{\substack{\emptyset\subseteq T\subseteq I_n \\ \|T\|=n}} \det(B_{n,T}) \\ &= \det(B_{n,I_n}) = \det(w_{i+j-1}),\end{aligned} \tag{5.8}$$

which is (5.2).

The equality of (5.2) and (5.3) is immediate. □

The elliptic function parameter relations are given by the following lemma.

Lemma 5.2. *Let* $z := 2K/\pi$, *as in* (2.1), *with* K *given by* (2.2) *and* k *the modulus. Take* q *as in* (2.4). *Let the classical theta functions* $\vartheta_3(0, q)$ *and* $\vartheta_2(0, q)$ *be defined by* (1.1) *and*

(1.2), *respectively. Then,*

$$z = \vartheta_3(0,q)^2, \tag{5.9}$$

$$z\sqrt{1-k^2} = \vartheta_3(0,-q)^2 = \vartheta_4(0,q)^2, \tag{5.10}$$

$$zk = \vartheta_2(0,q)^2, \tag{5.11}$$

$$4z^2k = \vartheta_2(0,q^{1/2})^4, \tag{5.12}$$

$$z^2(1+k^2) = 1 + 24\sum_{r=1}^{\infty}\frac{rq^r}{1+q^r}, \tag{5.13}$$

$$z^2(2-k^2) = 2 + 24\sum_{r=1}^{\infty}\frac{2rq^{2r}}{1+q^{2r}}, \tag{5.14}$$

$$z^2(1-2k^2) = 1 - 24\sum_{r=1}^{\infty}\frac{(2r-1)q^{2r-1}}{1+q^{2r-1}}. \tag{5.15}$$

Proof: Essentially all of (5.9)–(5.15) can be found in [117, Eq. (4), Section 40; Eq. (6), Section 40; Eq. (5), Section 40; Eq. (7), Section 40]. For convenience, we give more recent references. For the fundamental (5.9) see either [250, pp. 499–500], [22, Eq. (7.10), p. 27], or [21, Entry 6, p. 101 (keep in mind Eq. (3.1) on p. 98 and Eq. (6.12) on p. 102)]. For (5.10) see [22, Theorem 8.1 (Eq. (ii)), p. 27] or [21, Entry 10(ii), p. 122]. Equation (5.11) is equivalent to the relation

$$\Delta(q^2) = \tfrac{1}{2}z^{1/2}(k^2/q)^{1/4} \tag{5.16}$$

in [22, Theorem 8.2 (Eq. (iii)), p. 28] or [21, Entry 11(iii), p. 123], where $\Delta(q)$ from [22, Eq. (1.5), p. 3] is defined by

$$\Delta(q) := \sum_{j=0}^{\infty} q^{\frac{1}{2}j(j+1)}. \tag{5.17}$$

To see this, take the square root of both sides of (5.11), apply the relation

$$\vartheta_2(0,q) = 2q^{1/4}\Delta(q^2), \tag{5.18}$$

and simplify. Similarly, Eq. (5.12) is equivalent to the relation

$$\Delta(q) = 2^{-1/2}z^{1/2}k^{1/4}q^{-1/8} \tag{5.19}$$

in [22, Theorem 8.2 (Eq. (i)), p. 28] or [21, Entry 11(i), p. 123]. Equation (5.13) appears in [21, Entry 13(viii), p. 127 (keep in mind Eq. (6.12) on p. 102)]. Equation (5.14) is immediate from multiplying by 2 both sides of the relation in [21, Entry 13(ix), p. 127 (keep in mind Eq. (6.12) on p. 102)]. Finally, Eq. (5.15) is an immediate consequence of subtracting (5.13) from (5.14). □

We are now ready to obtain the main identities of this paper. We begin with the single Hankel determinant form of the $4n^2$ squares identity in the following theorem.

Theorem 5.3. *Let $\vartheta_3(0,-q)$ be determined by* (1.1), *and let $n = 1, 2, 3, \ldots$. We then have*

$$\vartheta_3(0,-q)^{4n^2} = \left\{(-1)^n 2^{2n^2+n} \prod_{r=1}^{2n-1} (r!)^{-1}\right\} \cdot \det(g_{r+s-1})_{1\le r,s\le n}, \tag{5.20}$$

with

$$g_i := U_{2i-1} - c_i, \tag{5.21}$$

where U_{2i-1} and c_i are defined by

$$U_{2i-1} \equiv U_{2i-1}(q) := \sum_{r=1}^{\infty} (-1)^{r-1} \frac{r^{2i-1} q^r}{1+q^r}, \quad \textit{for } i = 1, 2, 3, \ldots, \tag{5.22}$$

$$c_i := (-1)^{i-1} \frac{(2^{2i}-1)}{4i} \cdot |B_{2i}|, \quad \textit{for } i = 1, 2, 3, \ldots, \tag{5.23}$$

with B_{2i} the Bernoulli numbers defined by (2.60).

Proof: Our analysis deals with $\mathrm{sc}(u,k)\,\mathrm{dn}(u,k)$. Starting with (2.80), applying row operations and (3.66), appealing to (4.15), and then utilizing (5.9) we have the following computation.

$$H_n^{(1)}\left(\left\{U_{2\nu-1}(-q) - (-1)^{\nu-1} \frac{(2^{2\nu}-1)}{4\nu} \cdot |B_{2\nu}|\right\}\right) \tag{5.24}$$

$$= H_n^{(1)}\left(\left\{(-1)^{\nu} \frac{z^{2\nu}}{2^{2\nu+1}} \cdot (sd/c)_\nu(k^2)\right\}\right) \tag{5.25}$$

$$= \left\{(-1)^n 2^{-(2n^2+n)} \prod_{r=1}^{2n-1} r!\right\} \cdot z^{2n^2} \tag{5.26}$$

$$= \left\{(-1)^n 2^{-(2n^2+n)} \prod_{r=1}^{2n-1} r!\right\} \cdot \vartheta_3(0,q)^{4n^2}. \tag{5.27}$$

Replacing q by $-q$ in (5.24) and (5.27) gives

$$H_n^{(1)}\left(\left\{U_{2\nu-1}(q) - (-1)^{\nu-1} \frac{(2^{2\nu}-1)}{4\nu} \cdot |B_{2\nu}|\right\}\right)$$
$$= \left\{(-1)^n 2^{-(2n^2+n)} \prod_{r=1}^{2n-1} r!\right\} \cdot \vartheta_3(0,-q)^{4n^2}. \tag{5.28}$$

Solving for $\vartheta_3(0,-q)^{4n^2}$ in (5.28) yields (5.20). □

Lemma 5.1 and Theorem 5.3 lead to the determinant sum form of the $4n^2$ squares identity in the following theorem.

Theorem 5.4. *Let $n = 1, 2, 3, \ldots$. Then*

$$\vartheta_3(0, -q)^{4n^2} = 1 + \sum_{p=1}^{n} (-1)^p 2^{2n^2+n} \prod_{r=1}^{2n-1} (r!)^{-1} \sum_{\substack{\emptyset \subset S \subseteq I_n \\ \|S\|=p}} \det(M_{n,S}), \tag{5.29}$$

where $\vartheta_3(0, -q)$ is determined by (1.1), *and $M_{n,S}$ is the $n \times n$ matrix whose i-th row is*

$$U_{2i-1}, U_{2(i+1)-1}, \ldots, U_{2(i+n-1)-1}, \quad \text{if } i \in S$$

and

$$c_i, c_{i+1}, \ldots, c_{i+n-1}, \quad \text{if } i \notin S, \tag{5.30}$$

where U_{2i-1} and c_i are defined by (5.22) *and* (5.23), *respectively, with B_{2i} the Bernoulli numbers defined by* (2.60).

Proof: Specialize v_i and w_i as follows in Eq. (5.3) of Lemma 5.1 and then utilize (2.80) to write $v_i + w_i$ as a Lambert series.

$$v_i = (-1)^{i-1}\frac{(2^{2i}-1)}{4i} \cdot |B_{2i}|, \quad \text{for } i = 1, 2, 3, \ldots, \tag{5.31}$$

$$w_i = (-1)^i \frac{z^{2i}}{2^{2i+1}} \cdot (sd/c)_i(k^2), \quad \text{for } i = 1, 2, 3, \ldots, \tag{5.32}$$

$$v_i + w_i = U_{2i-1}(-q) = \sum_{r=1}^{\infty} \frac{-r^{2i-1}q^r}{1+(-1)^r q^r}, \quad \text{for } i = 1, 2, 3, \ldots. \tag{5.33}$$

Equating (5.25) and (5.27) it is immediate that

$$\det(w_{i+j-1}) = \left\{(-1)^n 2^{-(2n^2+n)} \prod_{r=1}^{2n-1} r!\right\} \cdot \vartheta_3(0, q)^{4n^2}. \tag{5.34}$$

From (4.56) we have

$$\det(v_{i+j-1}) = 2^{-(2n^2+n)} \prod_{r=1}^{2n-1} r!. \tag{5.35}$$

Equation (5.29) is now a direct consequence of applying the determinant evaluations in (5.34) and (5.35), replacing q by $-q$, and then multiplying both sides of the resulting transformation of (5.3) by

$$(-1)^n 2^{2n^2+n} \prod_{r=1}^{2n-1} (r!)^{-1}, \tag{5.36}$$

and simplifying. □

The single Hankel determinant form of the $4n(n+1)$ squares identity is given by the following theorem.

Theorem 5.5. *Let $\vartheta_3(0,-q)$ be determined by* (1.1), *and let $n = 1, 2, 3, \ldots$. We then have*

$$\vartheta_3(0,-q)^{4n(n+1)} = \left\{2^{2n^2+3n}\prod_{r=1}^{2n}(r!)^{-1}\right\}\cdot \det(g_{r+s-1})_{1\le r,s\le n}, \tag{5.37}$$

with

$$g_i := G_{2i+1} - a_i, \tag{5.38}$$

where G_{2i+1} and a_i are defined by

$$G_{2i+1} \equiv G_{2i+1}(q) := \sum_{r=1}^{\infty}(-1)^r\frac{r^{2i+1}q^r}{1-q^r}, \quad \text{for } i = 1,2,3,\ldots, \tag{5.39}$$

$$a_i := (-1)^i\frac{(2^{2i+2}-1)}{4(i+1)}\cdot|B_{2i+2}|, \quad \text{for } i = 1,2,3,\ldots, \tag{5.40}$$

with B_{2i+2} the Bernoulli numbers defined by (2.60).

Proof: Our analysis deals with $\mathrm{sc}^2(u,k)\,\mathrm{dn}^2(u,k)$. Starting with (2.81), applying row operations and (3.66), appealing to (4.17), and then utilizing (5.9) we have the following computation.

$$H_n^{(1)}\left(\left\{G_{2\nu+1}(-q) - (-1)^\nu\frac{(2^{2\nu+2}-1)}{4(\nu+1)}\cdot|B_{2\nu+2}|\right\}\right) \tag{5.41}$$

$$= H_n^{(1)}\left(\left\{(-1)^{\nu-1}\frac{z^{2\nu+2}}{2^{2\nu+3}}\cdot(s^2d^2/c^2)_\nu(k^2)\right\}\right) \tag{5.42}$$

$$= \left\{2^{-(2n^2+3n)}\prod_{r=1}^{2n} r!\right\}\cdot z^{2n(n+1)} \tag{5.43}$$

$$= \left\{2^{-(2n^2+3n)}\prod_{r=1}^{2n} r!\right\}\cdot \vartheta_3(0,q)^{4n(n+1)}. \tag{5.44}$$

Replacing q by $-q$ in (5.41) and (5.44) gives

$$H_n^{(1)}\left(\left\{G_{2\nu+1}(q) - (-1)^\nu\frac{(2^{2\nu+2}-1)}{4(\nu+1)}\cdot|B_{2\nu+2}|\right\}\right)$$

$$= \left\{2^{-(2n^2+3n)}\prod_{r=1}^{2n} r!\right\}\cdot \vartheta_3(0,-q)^{4n(n+1)}. \tag{5.45}$$

Solving for $\vartheta_3(0,-q)^{4n(n+1)}$ in (5.45) yields (5.37). □

The determinant sum form of the $4n(n+1)$ squares identity is given by the following theorem.

Theorem 5.6. *Let $n = 1, 2, 3, \ldots$. Then*

$$\vartheta_3(0,-q)^{4n(n+1)} = 1 + \sum_{p=1}^{n} (-1)^{n-p} 2^{2n^2+3n} \prod_{r=1}^{2n} (r!)^{-1} \sum_{\substack{\emptyset \subset S \subseteq I_n \\ \|S\|=p}} \det(M_{n,S}), \tag{5.46}$$

where $\vartheta_3(0,-q)$ is determined by (1.1), *and $M_{n,S}$ is the $n \times n$ matrix whose i-th row is*

$$G_{2i+1}, G_{2(i+1)+1}, \ldots, G_{2(i+n-1)+1}, \quad \text{if } i \in S$$

and

$$a_i, a_{i+1}, \ldots, a_{i+n-1}, \text{ if } i \notin S, \tag{5.47}$$

where G_{2i+1} and a_i are defined by (5.39) *and* (5.40), *respectively, with B_{2i} the Bernoulli numbers defined by* (2.60).

Proof: Specialize v_i and w_i as follows in Eq. (5.3) of Lemma 5.1 and then utilize (2.81) to write $v_i + w_i$ as a Lambert series.

$$v_i = (-1)^i \frac{(2^{2i+2}-1)}{4(i+1)} \cdot |B_{2i+2}|, \quad \text{for } i = 1, 2, 3, \ldots, \tag{5.48}$$

$$w_i = (-1)^{i-1} \frac{z^{2i+2}}{2^{2i+3}} \cdot (s^2d^2/c^2)_i(k^2), \quad \text{for } i = 1, 2, 3, \ldots, \tag{5.49}$$

$$v_i + w_i = G_{2i+1}(-q) = \sum_{r=1}^{\infty} \frac{r^{2i+1} q^r}{1-(-1)^r q^r}, \quad \text{for } i = 1, 2, 3, \ldots. \tag{5.50}$$

Equating (5.42) and (5.44) it is immediate that

$$\det(w_{i+j-1}) = \left\{ 2^{-(2n^2+3n)} \prod_{r=1}^{2n} r! \right\} \cdot \vartheta_3(0,q)^{4n(n+1)}. \tag{5.51}$$

From (4.57) we have

$$\det(v_{i+j-1}) = (-1)^n 2^{-(2n^2+3n)} \prod_{r=1}^{2n} r!. \tag{5.52}$$

Equation (5.46) is now a direct consequence of applying the determinant evaluations in (5.51) and (5.52), replacing q by $-q$, and then multiplying both sides of the resulting transformation of (5.3) by

$$2^{2n^2+3n} \prod_{r=1}^{2n} (r!)^{-1}, \tag{5.53}$$

and simplifying. □

The analysis of the formulas for $r_{4n^2}(N)$ and $r_{4n(n+1)}(N)$ obtained by taking the coefficient of q^N in Theorems 5.4 and 5.6 is analogous to the formulas for $r_{16}(n)$ and $r_{24}(n)$ in Theorem 1.7. The dominate terms for $r_{4n^2}(N)$ and $r_{4n(n+1)}(N)$ arise from the $p = n$ terms in (5.29) and (5.46), respectively. The other terms are all of a strictly decreasing lower order of magnitude. That is, the terms for $r_{4n^2}(N)$ and $r_{4n(n+1)}(N)$ corresponding to the p-th terms in (5.29) and (5.46) have orders of magnitude $N^{(4np-2p^2-1)}$ and $N^{(4np-2p^2+2p-1)}$, respectively. The dominate $p = n$ cases are consistent with [94, Eq. (9.20), p. 122]. Note that this analysis does not apply to the $n = 1$ case of Theorem 5.4. All of this analysis depends upon Lemma 5.37 at the end of this section.

In order to state the next four identities related to $\mathrm{nc}(u, k)$ we need the Euler numbers E_n defined by (2.61).

We begin with the single $H_n^{(1)}$ Hankel determinant identity in the following theorem.

Theorem 5.7. *Let $\vartheta_3(0, -q) = \vartheta_4(0, q)$ be determined by* (1.1), *and let $n = 1, 2, 3, \ldots$. We then have*

$$\vartheta_3(0,q)^{2n(n-1)}\vartheta_3(0,-q)^{2n^2} = \vartheta_3(0,q)^{2n(n-1)}\vartheta_4(0,q)^{2n^2}$$
$$= \left\{(-1)^n 2^{2n}\prod_{r=1}^{n-1}(2r)!^{-2}\right\}\cdot \det(g_{r+s-1})_{1\leq r,s\leq n}, \tag{5.54}$$

with

$$g_i := R_{2i-2} - b_i, \tag{5.55}$$

where R_{2i-2} and b_i, with $i = 1, 2, 3, \ldots$, are defined by

$$R_{2i-2} \equiv R_{2i-2}(q) := \sum_{r=1}^{\infty}(-1)^{r+1}\frac{(2r-1)^{2i-2}q^{2r-1}}{1+q^{2r-1}}, \tag{5.56}$$

and

$$b_i := (-1)^{i-1}\cdot \tfrac{1}{4}\cdot |E_{2i-2}|, \tag{5.57}$$

with E_{2i-2} the Euler numbers defined by (2.61).

Proof: Our analysis deals with $\mathrm{nc}(u, k)$. Starting with (2.82), applying row operations and (3.66), appealing to (4.12), and then utilizing both (5.9) and (5.10) we have the following computation.

$$H_n^{(1)}(\{R_{2\nu-2}(q) - (-1)^{\nu-1}\cdot \tfrac{1}{4}\cdot |E_{2\nu-2}|\}) \tag{5.58}$$
$$= H_n^{(1)}\left(\left\{(-1)^\nu \frac{z^{2\nu-1}}{4}\sqrt{1-k^2}\cdot (nc)_{\nu-1}(k^2)\right\}\right) \tag{5.59}$$
$$= \left\{(-1)^n 2^{-2n}\prod_{r=1}^{n-1}(2r)!^2\right\}\cdot z^{2n^2-n}(\sqrt{1-k^2})^{n^2} \tag{5.60}$$

$$= \left\{(-1)^n 2^{-2n} \prod_{r=1}^{n-1} (2r)!^2\right\} \cdot z^{n(n-1)} (z\sqrt{1-k^2})^{n^2} \tag{5.61}$$

$$= \left\{(-1)^n 2^{-2n} \prod_{r=1}^{n-1} (2r)!^2\right\} \cdot \vartheta_3(0,q)^{2n(n-1)} \vartheta_3(0,-q)^{2n^2}. \tag{5.62}$$

Equating (5.58) with (5.62) and then solving for $\vartheta_3(0,q)^{2n(n-1)}\vartheta_3(0,-q)^{2n^2}$ yields (5.54).

□

Taking $q \mapsto -q$ in the $n = 1$ case of Theorem 5.7 gives Jacobi's 2-squares identity in [117, Eq. (4), Section 40] and [21, Entry 8(i), p. 114]. The $n = 2$ case of Theorem 5.7 immediately leads to

$$\vartheta_3(0,q)^4 \vartheta_3(0,-q)^8 = \tfrac{1}{4} \det \begin{vmatrix} 4R_0 - 1 & 4R_2 + 1 \\ 4R_2 + 1 & 4R_4 - 5 \end{vmatrix}, \tag{5.63}$$

where R_{2i-2} is defined by (5.56), and $\vartheta_3(0,-q) = \vartheta_4(0,q)$ is determined by (1.1). The relation in (5.63) is an elegant formula for the product of the two identities in [21, Entry 8(ii), p. 114; Examples (i)($q \mapsto -q$), p. 139]. It is also analogous to the Lambert series expansions in Section 40 of [117].

We generally want the powers of the theta functions in this paper to be quadratic in n. However, it is sometimes useful to write the left-hand side of (5.54) in the form

$$\vartheta_3(0,-q^2)^{4n(n-1)} \vartheta_3(0,-q)^{2n}, \tag{5.64}$$

by appealing to the relation

$$\vartheta_3(0,q)\vartheta_3(0,-q) = \vartheta_3(0,-q^2)^2 \tag{5.65}$$

from [21, Entry 25(iii), p. 40].

The $H_n^{(1)}$ Hankel determinant sum identity is given by the following theorem.

Theorem 5.8. *Let $n = 1, 2, 3, \ldots$. Then*

$$\begin{aligned} \vartheta_3(0,q)^{2n(n-1)} \vartheta_3(0,-q)^{2n^2} &= \vartheta_3(0,q)^{2n(n-1)} \vartheta_4(0,q)^{2n^2} \\ &= 1 + \sum_{p=1}^{n} (-1)^p 2^{2n} \prod_{r=1}^{n-1} (2r)!^{-2} \sum_{\substack{\emptyset \subset S \subseteq I_n \\ \|S\| = p}} \det(M_{n,S}), \end{aligned} \tag{5.66}$$

where $\vartheta_3(0,-q) = \vartheta_4(0,q)$ is determined by (1.1), *and $M_{n,S}$ is the $n \times n$ matrix whose i-th row is*

$$R_{2i-2}, R_{2(i+1)-2}, \ldots, R_{2(i+n-1)-2}, \quad \text{if } i \in S$$

and

$$b_i, b_{i+1}, \ldots, b_{i+n-1}, \quad \text{if } i \notin S, \tag{5.67}$$

where R_{2i-2} *and* b_i *are defined by* (5.56) *and* (5.57), *respectively, with* E_{2i-2} *the Euler numbers defined by* (2.61).

Proof: Specialize v_i and w_i as follows in Eq. (5.3) of Lemma 5.1 and then utilize (2.82) to write $v_i + w_i$ as a Lambert series.

$$v_i = (-1)^{i-1} \cdot \tfrac{1}{4} \cdot |E_{2i-2}|, \tag{5.68}$$

$$w_i = (-1)^i \frac{z^{2i-1}}{4} \sqrt{1-k^2} \cdot (nc)_{i-1}(k^2), \tag{5.69}$$

$$v_i + w_i = R_{2i-2}(q) = \sum_{r=1}^{\infty} (-1)^{r+1} \frac{(2r-1)^{2i-2} q^{2r-1}}{1+q^{2r-1}}, \tag{5.70}$$

for $i = 1, 2, 3, \ldots$.

Equating (5.59) and (5.62) it is immediate that

$$\det(w_{i+j-1}) = \left\{ (-1)^n 2^{-2n} \prod_{r=1}^{n-1} (2r)!^2 \right\} \cdot \vartheta_3(0,q)^{2n(n-1)} \vartheta_3(0,-q)^{2n^2}. \tag{5.71}$$

From (4.58) we have

$$\det(v_{i+j-1}) = 2^{-2n} \prod_{r=1}^{n-1} (2r)!^2. \tag{5.72}$$

Equation (5.66) is now a direct consequence of applying the determinant evaluations in (5.71) and (5.72), and then multiplying both sides of the resulting transformation of (5.3) by

$$(-1)^n 2^{2n} \prod_{r=1}^{n-1} (2r)!^{-2}, \tag{5.73}$$

and simplifying. □

Taking $q \mapsto -q$ in the $n = 1$ case of Theorem 5.8 gives Jacobi's 2-squares identity in [117, Eq. (4), Section 40] and [21, Entry 8(i), p. 114]. The $n = 2$ case of Theorem 5.8 immediately leads to

$$\vartheta_3(0,q)^4 \vartheta_3(0,-q)^8 = 1 - [5R_0 + 2R_2 + R_4] + 4\left[R_0 R_4 - R_2^2\right], \tag{5.74}$$

where R_{2i-2} is defined by (5.56), and $\vartheta_3(0,-q) = \vartheta_4(0,q)$ is determined by (1.1).

The single $H_n^{(2)}$ Hankel determinant identity is given by the following theorem.

Theorem 5.9. *Let $\vartheta_3(0,-q) = \vartheta_4(0,q)$ be determined by (1.1), and let $n = 1, 2, 3, \ldots$. We then have*

$$\begin{aligned}\vartheta_3(0,q)^{2n(n+1)}\vartheta_3(0,-q)^{2n^2} &= \vartheta_3(0,q)^{2n(n+1)}\vartheta_4(0,q)^{2n^2} \\ &= \left\{2^{2n}\prod_{r=1}^{n}(2r-1)!^{-2}\right\}\cdot \det(g_{r+s-1})_{1\le r,s\le n}, \end{aligned} \tag{5.75}$$

with

$$g_i := R_{2i} - b_{i+1}, \tag{5.76}$$

where R_{2i} and b_{i+1} are determined by (5.56) and (5.57), respectively, with E_{2i} the Euler numbers defined by (2.61).

Proof: Our analysis still deals with $\mathrm{nc}(u,k)$. Starting with (2.82), applying row operations and (3.66), appealing to (4.24), and then utilizing both (5.9) and (5.10) we have the following computation.

$$H_n^{(2)}\left(\left\{R_{2\nu-2}(q) - (-1)^{\nu-1}\cdot\tfrac{1}{4}\cdot|E_{2\nu-2}|\right\}\right) \tag{5.77}$$

$$= H_n^{(1)}\left(\left\{R_{2\nu}(q) - (-1)^{\nu}\cdot\tfrac{1}{4}\cdot|E_{2\nu}|\right\}\right) \tag{5.78}$$

$$= H_n^{(1)}\left(\left\{(-1)^{\nu+1}\frac{z^{2\nu+1}}{4}\sqrt{1-k^2}\cdot(nc)_\nu(k^2)\right\}\right) \tag{5.79}$$

$$= \left\{2^{-2n}\prod_{r=1}^{n}(2r-1)!^2\right\}\cdot z^{2n^2+n}(\sqrt{1-k^2})^{n^2} \tag{5.80}$$

$$= \left\{2^{-2n}\prod_{r=1}^{n}(2r-1)!^2\right\}\cdot z^{n(n+1)}(z\sqrt{1-k^2})^{n^2} \tag{5.81}$$

$$= \left\{2^{-2n}\prod_{r=1}^{n}(2r-1)!^2\right\}\cdot \vartheta_3(0,q)^{2n(n+1)}\vartheta_3(0,-q)^{2n^2}. \tag{5.82}$$

Equating (5.78) with (5.82) and then solving for $\vartheta_3(0,q)^{2n(n+1)}\vartheta_3(0,-q)^{2n^2}$ yields (5.75). □

The $n = 1$ case of Theorem 5.9 is

$$\vartheta_3(0,q)^4\vartheta_3(0,-q)^2 = 1 + 4R_2 = 1 - 4\sum_{r=1}^{\infty}(-1)^r\frac{(2r-1)^2q^{2r-1}}{1+q^{2r-1}}, \tag{5.83}$$

which is Jacobi's degree 6 Lambert series expansion in [117, Eq. (41), Section 40]. Note that (5.83) gives an elegant formula for the product of the two identities in [21, Entry 8(ii), p. 114; Entry 8(v), p. 114].

The $n = 2$ case of Theorem 5.9 is

$$\vartheta_3(0,q)^{12}\vartheta_3(0,-q)^8 = \frac{1}{(3!)^2}\det\begin{vmatrix} 4R_2+1 & 4R_4-5 \\ 4R_4-5 & 4R_6+61 \end{vmatrix}, \tag{5.84}$$

where R_{2i} is determined by (5.56), and $\vartheta_3(0,-q) = \vartheta_4(0,q)$ is determined by (1.1).

The corresponding $H_n^{(2)}$ Hankel determinant sum identity is given by the following theorem.

Theorem 5.10. *Let $n = 1, 2, 3, \ldots$. Then*

$$\vartheta_3(0,q)^{2n(n+1)}\vartheta_3(0,-q)^{2n^2} = \vartheta_3(0,q)^{2n(n+1)}\vartheta_4(0,q)^{2n^2}$$
$$= 1 + \sum_{p=1}^{n}(-1)^{n-p}2^{2n}\prod_{r=1}^{n}(2r-1)!^{-2}\sum_{\substack{\emptyset\subset S\subseteq I_n\\ \|S\|=p}}\det(M_{n,S}), \tag{5.85}$$

where $\vartheta_3(0,-q) = \vartheta_4(0,q)$ is determined by (1.1), *and $M_{n,S}$ is the $n \times n$ matrix whose i-th row is*

$$R_{2i}, R_{2(i+1)}, \ldots, R_{2(i+n-1)}, \quad \textit{if } i \in S$$

and

$$b_{i+1}, b_{i+2}, \ldots, b_{i+n}, \quad \textit{if } i \notin S, \tag{5.86}$$

where R_{2i} and b_{i+1} are determined by (5.56) *and* (5.57), *respectively, with E_{2i} the Euler numbers defined by* (2.61).

Proof: Specialize v_i and w_i as follows in Eq. (5.3) of Lemma 5.1 and then utilize (2.82) to write $v_i + w_i$ as a Lambert series.

$$v_i = (-1)^i \cdot \tfrac{1}{4} \cdot |E_{2i}|, \tag{5.87}$$
$$w_i = (-1)^{i+1}\frac{z^{2i+1}}{4}\sqrt{1-k^2}\cdot(nc)_i(k^2), \tag{5.88}$$
$$v_i + w_i = R_{2i}(q) = \sum_{r=1}^{\infty}(-1)^{r+1}\frac{(2r-1)^{2i}q^{2r-1}}{1+q^{2r-1}}, \tag{5.89}$$

for $i = 1, 2, 3, \ldots$.

Equating (5.79) and (5.82) it is immediate that

$$\det(w_{i+j-1}) = \left\{2^{-2n}\prod_{r=1}^{n}(2r-1)!^2\right\}\cdot\vartheta_3(0,q)^{2n(n+1)}\vartheta_3(0,-q)^{2n^2}. \tag{5.90}$$

From (4.59) we have

$$\det(v_{i+j-1}) = (-1)^n 2^{-2n} \prod_{r=1}^{n} (2r-1)!^2. \tag{5.91}$$

Equation (5.85) is now a direct consequence of applying the determinant evaluations in (5.90) and (5.91), and then multiplying both sides of the resulting transformation of (5.3) by

$$2^{2n} \prod_{r=1}^{n} (2r-1)!^{-2}, \tag{5.92}$$

and simplifying. □

Keeping in mind (2.90), the Maclaurin series expansions for $\mathrm{nc}(u,k)$ and $\mathrm{sn}(u,k)\,\mathrm{dn}(u,k)/\mathrm{cn}^2(u,k)$ in (2.66) and (2.79), and the derivative formula for $\mathrm{nc}(u,k)$, it follows that Theorem 5.9 is equivalent to the single $H_n^{(1)}$ Hankel determinant theta function identity related to $\mathrm{sn}(u,k)\,\mathrm{dn}(u,k)/\mathrm{cn}^2(u,k)$.

The next two single $H_n^{(1)}$ Hankel determinant identities involve the $z=0$ case of the theta function $\vartheta_2(z,q)$ in [250, p. 464], defined by (1.2). These identities are related to $\mathrm{sd}(u,k)\,\mathrm{cn}(u,k)$ and $\mathrm{sn}^2(u,k)$, respectively. They are the first major step in our proof of the two Kac–Wakimoto conjectured identities for triangular numbers in [120, p. 452].

We have the following theorem.

Theorem 5.11. *Let $\vartheta_2(0,q)$ be defined by* (1.2), *and let $n = 1, 2, 3, \ldots$. We then have*

$$\vartheta_2(0,q)^{4n^2} = \left\{ 4^{n(n+1)} \prod_{r=1}^{2n-1} (r!)^{-1} \right\} \cdot \det\left(C_{2(r+s-1)-1}\right)_{1\le r,s\le n}, \tag{5.93}$$

and

$$\vartheta_2(0,q^{1/2})^{4n(n+1)} = \left\{ 2^{n(4n+5)} \prod_{r=1}^{2n} (r!)^{-1} \right\} \cdot \det\left(D_{2(r+s-1)+1}\right)_{1\le r,s\le n}, \tag{5.94}$$

where C_{2i-1} and D_{2i+1} are defined by, respectively,

$$C_{2i-1} \equiv C_{2i-1}(q) := \sum_{r=1}^{\infty} \frac{(2r-1)^{2i-1} q^{2r-1}}{1-q^{2(2r-1)}}, \quad \text{for } i = 1, 2, 3, \ldots, \tag{5.95}$$

$$D_{2i+1} \equiv D_{2i+1}(q) := \sum_{r=1}^{\infty} \frac{r^{2i+1} q^r}{1-q^{2r}}, \quad \text{for } i = 1, 2, 3, \ldots. \tag{5.96}$$

Proof: The proof of the first identity deals with $\mathrm{sd}(u,k)\,\mathrm{cn}(u,k)$. Starting with (2.83), applying row operations and (3.66), appealing to (4.9), and then utilizing (5.11) we have the following computation.

$$H_n^{(1)}(\{C_{2\nu-1}(q)\}) \tag{5.97}$$

$$= H_n^{(1)}\left(\left\{(-1)^{\nu-1}\frac{z^{2\nu}k^2}{2^{2\nu+2}}\cdot(sc/d)_\nu(k^2)\right\}\right) \tag{5.98}$$

$$= \left\{4^{-n(n+1)}\prod_{r=1}^{2n-1} r!\right\}\cdot(zk)^{2n^2} \tag{5.99}$$

$$= \left\{4^{-n(n+1)}\prod_{r=1}^{2n-1} r!\right\}\cdot\vartheta_2(0,q)^{4n^2}. \tag{5.100}$$

Equating (5.97) with (5.100) and then solving for $\vartheta_2(0,q)^{4n^2}$ yields (5.93).

The proof of the second identity deals with $\mathrm{sn}^2(u,k)$. Starting with (2.84), applying row operations and (3.66), appealing to (4.7), and then utilizing (5.12) we have the following computation.

$$H_n^{(1)}(\{D_{2\nu+1}(q)\}) \tag{5.101}$$

$$= H_n^{(1)}\left(\left\{(-1)^{\nu-1}\frac{z^{2\nu+2}k^2}{2^{2\nu+3}}\cdot(sn^2)_\nu(k^2)\right\}\right) \tag{5.102}$$

$$= \left\{2^{-n(4n+5)}\prod_{r=1}^{2n} r!\right\}\cdot(4z^2k)^{n(n+1)} \tag{5.103}$$

$$= \left\{2^{-n(4n+5)}\prod_{r=1}^{2n} r!\right\}\cdot\vartheta_2(0,q^{1/2})^{4n(n+1)}. \tag{5.104}$$

Equating (5.101) with (5.104) and then solving for $\vartheta_2(0,q^{1/2})^{4n(n+1)}$ yields (5.94). □

It is sometimes useful to rewrite (5.94) by observing from (5.12), (5.11), and (5.9) that

$$\vartheta_2(0,q^{1/2})^4 = 4(zk)z = 4\vartheta_2(0,q)^2\vartheta_3(0,q)^2. \tag{5.105}$$

Equation (5.94) then becomes

$$[\vartheta_2(0,q)\vartheta_3(0,q)]^{2n(n+1)} = \left\{2^{n(2n+3)}\prod_{r=1}^{2n}(r!)^{-1}\right\}\cdot\det\left(D_{2(r+s-1)+1}\right)_{1\le r,s\le n}, \tag{5.106}$$

where D_{2i+1} is defined by (5.96).

The $n=1$ case of Eq. (5.93) of Theorem 5.11, written as in (5.99), is given by [117, Eq. (9), Section 40]. Similarly, the $n=1$ case of (5.94), expressed as in (5.103), is given by [117, Eq. (3), Section 41]. Moreover, taking $q\mapsto -q$ or $q\mapsto q^2$ in the $n=1$ case of (5.103) yields Eqs. (4) or (5), respectively, of [117, Section 41].

Keeping in mind the Kac–Wakimoto conjectured identities for triangular numbers in [120, p. 452], we next rewrite Theorem 5.11 in terms of the sum $\Delta(q)$ in (5.17). Applying the relation (5.18) to the left-hand sides of (5.93) and (5.94) immediately gives the following corollary.

Corollary 5.12. *Let $\Delta(q)$ be defined by* (5.17), *and let $n = 1, 2, 3, \ldots$. We then have*

$$\Delta(q^2)^{4n^2} = \left\{4^{-n(n-1)}q^{-n^2}\prod_{r=1}^{2n-1}(r!)^{-1}\right\}\cdot\det\left(C_{2(r+s-1)-1}\right)_{1\le r,s\le n}, \tag{5.107}$$

and

$$\Delta(q)^{4n(n+1)} = \left\{2^n q^{-n(n+1)/2}\prod_{r=1}^{2n}(r!)^{-1}\right\}\cdot\det\left(D_{2(r+s-1)+1}\right)_{1\le r,s\le n}, \tag{5.108}$$

where C_{2i-1} and D_{2i+1} are defined by (5.95) *and* (5.96), *respectively.*

The $n = 1$ cases of (5.107) and (5.108) are the classical identities of Legendre [139], [21, Eqs. (ii) and (iii), p. 139] given by Theorem 7.7.

We next have the four single $H_n^{(1)}$ and $H_n^{(2)}$ Hankel determinant identities related to $\mathrm{cn}(u, k)$ and $\mathrm{dn}(u, k)$.

We begin with the following $H_n^{(1)}$ theorem.

Theorem 5.13. *Let $\vartheta_2(0, q)$ and $\vartheta_3(0, q)$ be defined by* (1.2) *and* (1.1), *respectively. Let $n = 1, 2, 3, \ldots$. We then have*

$$\vartheta_2(0,q)^{2n^2}\vartheta_3(0,q)^{2n(n-1)} = \left\{4^n\prod_{r=1}^{n-1}(2r)!^{-2}\right\}\cdot\det\left(T_{2(r+s-1)-2}\right)_{1\le r,s\le n}, \tag{5.109}$$

where T_{2i-2} is defined by

$$T_{2i-2} \equiv T_{2i-2}(q) := \sum_{r=1}^{\infty}\frac{(2r-1)^{2i-2}q^{r-\frac{1}{2}}}{1+q^{2r-1}}, \quad \textit{for } i = 1, 2, 3, \ldots. \tag{5.110}$$

Proof: Our analysis deals with $\mathrm{cn}(u, k)$. Starting with (2.85), applying row operations and (3.67), appealing to (4.2), and then utilizing both (5.9) and (5.11) we have the following computation.

$$H_n^{(1)}(\{T_{2\nu-2}(q)\}) \tag{5.111}$$

$$= H_n^{(1)}\left(\left\{(-1)^{\nu-1}\frac{z^{2\nu-1}k}{4}\cdot(cn)_{\nu-1}(k^2)\right\}\right) \tag{5.112}$$

$$= \left\{4^{-n}\prod_{r=1}^{n-1}(2r)!^2\right\}\cdot z^{2n^2-n}k^{n^2} \tag{5.113}$$

$$= \left\{4^{-n}\prod_{r=1}^{n-1}(2r)!^2\right\}\cdot(zk)^{n^2}z^{n(n-1)} \tag{5.114}$$

$$= \left\{4^{-n}\prod_{r=1}^{n-1}(2r)!^2\right\}\cdot\vartheta_2(0,q)^{2n^2}\vartheta_3(0,q)^{2n(n-1)}. \tag{5.115}$$

Equating (5.111) with (5.115) and then solving for $\vartheta_2(0,q)^{2n^2}\vartheta_3(0,q)^{2n(n-1)}$ yields (5.109). □

The $n=1$ case of Theorem 5.13, when written as in (5.114), is given by [117, Eq. (5), Section 40]. The $n=1$ case of (5.109) gives the identity in [21, Example (iv), p. 139].

Applying the relation (5.105) to the left-hand-side of (5.109) immediately implies that

$$\vartheta_2(0,q)^{2n}\vartheta_2(0,q^{1/2})^{4n(n-1)} = \left\{4^{n^2}\prod_{r=1}^{n-1}(2r)!^{-2}\right\}\cdot \det\left(T_{2(r+s-1)-2}\right)_{1\le r,s\le n}, \quad (5.116)$$

where T_{2i-2} is defined by (5.110). The form of the exponents and arguments in the left-hand-side of (5.116) is much closer than (5.109) to that of the product sides of the eta function identities in Appendix I of [151].

Applying (5.18) to the left-hand-side of the $n = 2$ case of (5.116) yields an elegant product formula for $\Delta(q^2)^4\Delta(q)^8$. This expresses the product of the sums in entries (ii) and (iii) of [21, p. 139] as a simple 2×2 determinant of Lambert series.

The single $H_n^{(2)}$ Hankel determinant identity related to $\mathrm{cn}(u,k)$ is given by the following theorem.

Theorem 5.14. *Let $\vartheta_2(0,q)$ and $\vartheta_3(0,q)$ be defined by* (1.2) *and* (1.1), *respectively. Let $n=1,2,3,\ldots$. We then have*

$$\vartheta_2(0,q)^{2n^2}\vartheta_3(0,q)^{2n(n+1)} = \left\{4^{n}\prod_{r=1}^{n}(2r-1)!^{-2}\right\}\cdot \det\left(T_{2(r+s-1)}\right)_{1\le r,s\le n}, \quad (5.117)$$

where T_{2i} is determined by (5.110).

Proof: Our analysis again deals with $\mathrm{cn}(u,k)$. Starting with (2.85), applying row operations and (3.66), appealing to (4.19), and then utilizing both (5.9) and (5.11) we have the following computation.

$$H_n^{(2)}(\{T_{2\nu-2}(q)\}) \quad (5.118)$$
$$= H_n^{(1)}(\{T_{2\nu}(q)\}) \quad (5.119)$$
$$= H_n^{(1)}\left(\left\{(-1)^{\nu}\frac{z^{2\nu+1}k}{4}\cdot(cn)_{\nu}(k^2)\right\}\right) \quad (5.120)$$
$$= \left\{4^{-n}\prod_{r=1}^{n}(2r-1)!^2\right\}\cdot z^{2n^2+n}k^{n^2} \quad (5.121)$$
$$= \left\{4^{-n}\prod_{r=1}^{n}(2r-1)!^2\right\}\cdot (zk)^{n^2}z^{n(n+1)} \quad (5.122)$$
$$= \left\{4^{-n}\prod_{r=1}^{n}(2r-1)!^2\right\}\cdot \vartheta_2(0,q)^{2n^2}\vartheta_3(0,q)^{2n(n+1)}. \quad (5.123)$$

Equating (5.119) with (5.123) and then solving for $\vartheta_2(0,q)^{2n^2}\vartheta_3(0,q)^{2n(n+1)}$ yields (5.117). □

The $n=1$ case of Theorem 5.14, when written as in (5.121), is given by [117, Eq. (40), Section 40].

Keeping in mind (2.88), the Maclaurin series expansions for $\mathrm{cn}(u,k)$ and $\mathrm{sn}(u,k)\,\mathrm{dn}(u,k)$ in (2.62) and (2.74), and the derivative formula for $\mathrm{cn}(u,k)$, it follows that Theorem 5.14 is equivalent to the single $H_n^{(1)}$ Hankel determinant theta function identity related to $\mathrm{sn}(u,k)$ $\mathrm{dn}(u,k)$.

In [185, p. 331] Perron equates the Laplace transform of $\mathrm{cn}(u,k)$ with its C-fraction expansion and Fourier series. This was just a convenient reference. No applications of this equality of all three was given.

Applying the relation (5.105) to the left-hand-side of (5.117) immediately implies that

$$\vartheta_2(0,q^{1/2})^{4n^2}\vartheta_3(0,q)^{2n} = \left\{4^{n(n+1)}\prod_{r=1}^{n}(2r-1)!^{-2}\right\}\cdot\det\left(T_{2(r+s-1)}\right)_{1\le r,s\le n}, \tag{5.124}$$

where T_{2i} is determined by (5.110).

We next utilize (5.18), with $q\mapsto q^{1/2}$, to rewrite the left-hand-side of (5.124). Some further simplification then yields the following corollary.

Corollary 5.15. *Let $\vartheta_3(0,q)$ and $\Delta(q)$ be defined by* (1.1) *and* (5.17), *respectively. Let $n=1,2,3,\ldots$. We then have*

$$\vartheta_3(0,q)^{2n}\Delta(q)^{(2n)^2} = \left\{4^{-n(n-1)}q^{-n(n-1)/2}\prod_{r=1}^{n}(2r-1)!^{-2}\right\}\cdot\det\left(\tilde{T}_{2(r+s-1)}\right)_{1\le r,s\le n}, \tag{5.125}$$

where $\tilde{T}_{2i}$ is defined by

$$\tilde{T}_{2i}\equiv\tilde{T}_{2i}(q) := \sum_{r=1}^{\infty}\frac{(2r-1)^{2i}q^{r-1}}{1+q^{2r-1}},\quad \text{for } i=1,2,3,\ldots. \tag{5.126}$$

When viewed as a generating function in q^N, (for $N=0,1,2,3,\ldots$), the left-hand-side of (5.125) has an elegant combinatorial interpretation. Motivated by [6, pp. 506–508], the coefficient of q^N is the number of ways of writing N as a sum of $2n$ squares and $(2n)^2$ triangular numbers. The resulting expansion on the right-hand-side of (5.125) is just as simple as those in Corollary 5.12.

The $n=1$ case of Corollary 5.15 is given by [21, Example (v), p. 139].

We now consider the theta function identities related to $\mathrm{dn}(u,k)$. We start with the following $H_n^{(1)}$ theorem.

Theorem 5.16. *Let $\vartheta_2(0,q)$ and $\vartheta_3(0,q)$ be defined by* (1.2) *and* (1.1), *respectively. Let $n=1,2,3,\ldots$. We then have*

$$\vartheta_2(0,q)^{2n(n-1)}\vartheta_3(0,q)^{2n^2} = \left\{4^{n^2}\prod_{r=1}^{n-1}(2r)!^{-2}\right\}\cdot\det\left(N_{2(r+s-1)-2}\right)_{1\le r,s\le n} \tag{5.127a}$$

$$+\left\{4^{n^2-1}\prod_{r=1}^{n-1}(2r)!^{-2}\right\}\cdot\det\left(N_{2(r+s)}\right)_{1\le r,s\le n-1}, \tag{5.127b}$$

where N_{2i-2} is defined by

$$N_{2i-2} \equiv N_{2i-2}(q) := \sum_{r=1}^{\infty} \frac{r^{2i-2}q^r}{1+q^{2r}}, \quad \textit{for } i = 1, 2, 3, \ldots. \tag{5.128}$$

Proof: Our analysis deals with $\mathrm{dn}(u, k)$. Starting with (2.86) and (2.87), solving for $(-1)^m(z^{2m+1}/2^{2m+2}) \cdot (dn)_m(k^2)$ for $m \geq 0$, applying row operations, recalling (3.55), and simplifying, leads to the following identity.

$$H_n^{(1)}(\{N_{2\nu-2}(q)\}) + \tfrac{1}{4}H_{n-1}^{(3)}(\{N_{2\nu-2}(q)\}) \tag{5.129}$$

$$= H_n^{(0)}\left(\left\{(-1)^\nu \frac{z^{2\nu+1}}{4^{\nu+1}} \cdot (dn)_\nu(k^2)\right\}\right)$$

$$= H_n^{(1)}\left(\left\{(-1)^{\nu-1} \frac{z^{2\nu-1}}{4^\nu} \cdot (dn)_{\nu-1}(k^2)\right\}\right). \tag{5.130}$$

Next, applying row operations and (3.67) to (5.130), appealing to (4.2), and then utilizing both (5.9) and (5.11) we obtain

$$\left\{4^{-n^2}\prod_{r=1}^{n-1}(2r)!^2\right\} \cdot z^{2n^2-n}k^{n^2-n} \tag{5.131}$$

$$= \left\{4^{-n^2}\prod_{r=1}^{n-1}(2r)!^2\right\} \cdot (zk)^{n(n-1)}z^{n^2} \tag{5.132}$$

$$= \left\{4^{-n^2}\prod_{r=1}^{n-1}(2r)!^2\right\} \cdot \vartheta_2(0,q)^{2n(n-1)}\vartheta_3(0,q)^{2n^2}. \tag{5.133}$$

Equating (5.129) with (5.133) and then solving for $\vartheta_2(0,q)^{2n(n-1)}\vartheta_3(0,q)^{2n^2}$ yields (5.127). □

The $n = 1$ case of Theorem 5.16, when written as in (5.131), is given by [117, Eq. (4), Section 40].

The $n = 1$ case of Theorem 5.16 is equivalent to Jacobi's 2-squares identity in [21, Entry 8(i), p. 114]. To see this, note that the 0×0 determinant in (5.127b) is defined to be 1, and that an elementary argument [21, p. 115] involving geometric series and an interchange of summation gives

$$\sum_{r=1}^{\infty} \frac{q^r}{1+q^{2r}} = \sum_{r=1}^{\infty} \frac{(-1)^{r-1}q^{2r-1}}{1-q^{2r-1}}. \tag{5.134}$$

Multiplying both sides of (5.127) by 4^{n^2}, it is immediate from (5.105) that the product side of (5.127) can be written as $\vartheta_2(0,q^{1/2})^{4n^2}\vartheta_2(0,q)^{-2n}$. Just as in (5.116), this form is much closer to [151, Appendix I].

The single $H_n^{(2)}$ Hankel determinant identity related to $\mathrm{dn}(u,k)$ is given by the following theorem.

Theorem 5.17. *Let $\vartheta_2(0,q)$ and $\vartheta_3(0,q)$ be defined by* (1.2) *and* (1.1), *respectively. Let $n = 1, 2, 3, \ldots$. We then have*

$$\vartheta_2(0,q)^{2n(n+1)}\vartheta_3(0,q)^{2n^2} = \left\{4^{n(n+1)}\prod_{r=1}^{n}(2r-1)!^{-2}\right\}\cdot \det\left(N_{2(r+s-1)}\right)_{1\le r,s\le n}, \quad (5.135)$$

where N_{2i} is determined by (5.128).

Proof: Our analysis again deals with $\mathrm{dn}(u,k)$. Starting with (2.86), applying row operations and (3.66), appealing to (4.20), and then utilizing both (5.9) and (5.11) we have the following computation.

$$H_n^{(2)}(\{N_{2\nu-2}(q)\}) \quad (5.136)$$

$$= H_n^{(1)}(\{N_{2\nu}(q)\}) \quad (5.137)$$

$$= H_n^{(1)}\left(\left\{(-1)^\nu \frac{z^{2\nu+1}}{4^{\nu+1}}\cdot (dn)_\nu(k^2)\right\}\right) \quad (5.138)$$

$$= \left\{4^{-n(n+1)}\prod_{r=1}^{n}(2r-1)!^2\right\}\cdot z^{2n^2+n}k^{n^2+n} \quad (5.139)$$

$$= \left\{4^{-n(n+1)}\prod_{r=1}^{n}(2r-1)!^2\right\}\cdot (zk)^{n(n+1)}z^{n^2} \quad (5.140)$$

$$= \left\{4^{-n(n+1)}\prod_{r=1}^{n}(2r-1)!^2\right\}\vartheta_2(0,q)^{2n(n+1)}\vartheta_3(0,q)^{2n^2}. \quad (5.141)$$

Equating (5.137) with (5.141) and then solving for $\vartheta_2(0,q)^{2n(n+1)}\vartheta_3(0,q)^{2n^2}$ yields (5.135). □

The $n = 1$ case of Theorem 5.17, when written as in (5.139), is given by [117, Eq. (42), Section 40]. See also [21, Entry 17(ii), p. 138].

Keeping in mind (2.89), the Maclaurin series expansions for $\mathrm{dn}(u,k)$ and $\mathrm{sn}(u,k)\,\mathrm{cn}(u,k)$ in (2.63) and (2.75), and the derivative formula for $\mathrm{dn}(u,k)$, it follows that Theorem 5.17 is equivalent to the single $H_n^{(1)}$ Hankel determinant theta function identity related to $\mathrm{sn}(u,k)$ $\mathrm{cn}(u,k)$.

Applying the relation (5.105) to the left-hand-side of (5.135) immediately implies that

$$\vartheta_2(0,q^{1/2})^{4n^2}\vartheta_2(0,q)^{2n} = \left\{4^{n(2n+1)}\prod_{r=1}^{n}(2r-1)!^{-2}\right\}\cdot \det\left(N_{2(r+s-1)}\right)_{1\le r,s\le n}, \quad (5.142)$$

where N_{2i} is determined by (5.128).

We next utilize (5.18), with $q \mapsto q^{1/2}$, to rewrite the left-hand-side of (5.142). Some further simplification then yields the following corollary.

Corollary 5.18. *Let $\Delta(q)$ be defined by* (5.17). *Let $n = 1, 2, 3, \ldots$. We then have*

$$\Delta(q)^{(2n)^2}\Delta(q^2)^{2n} = \left\{q^{-n(n+1)/2}\prod_{r=1}^{n}(2r-1)!^{-2}\right\}\cdot\det\left(N_{2(r+s-1)}\right)_{1\leq r,s\leq n}, \qquad (5.143)$$

where N_{2i} is determined by (5.128).

Analogous to (5.125), the coefficient of q^N in (5.143) is the number of ways of writing N as a sum of $(2n)^2$ triangular numbers and $2n$ doubles of triangular numbers.

By considering quotients of our theta function identities related to a given Jacobi elliptic function and its derivative, or equivalently, to the pairs of $H_n^{(1)}$ and $H_n^{(2)}$ identities, we are able to write $\vartheta_2(0,q)^{4n}$ and $\vartheta_3(0,q)^{4n}$ as a quotient of $n \times n$ determinants of Lambert series. To obtain such an identity for $\vartheta_2(0,q)^{4n}$, divide (5.135) by (5.127). (Here, the denominator will involve a sum of two determinants.) The quotient of determinants identities for $\vartheta_3(0,q)^{4n}$ are simpler. The first two identities for $\vartheta_3(0,q)^{4n}$ are obtained by either dividing (5.75) by (5.54), or (5.117) by (5.109). The third results by dividing the $q \mapsto -q$ cases of (5.37) by (5.20). Equating pairs of these three quotients of determinants gives three determinant identities. Finally, equating the $n \mapsto n^2$ case of each of the three quotient of determinants identities for $\vartheta_3(0,q)^{4n}$ with the $q \mapsto -q$ case of (5.20) yields three more interesting determinantal identities involving Lambert series. Each of these last three identities equates a quotient of $n^2 \times n^2$ determinants to the product of an $n \times n$ determinant and a certain constant.

For convenience and future reference we now extract from the above analysis in this section our generalization to infinite families of all 21 of Jacobi's [117, Sections 40–42] explicitly stated degree 2, 4, 6, 8 Lambert series expansions of classical theta functions. We follow Jacobi in expressing everything in terms of the elliptic function parameters z, k, and q. For three of our identities we need the additional Lambert series defined by

$$\hat{C}_{2m-1}(q) := \sum_{r=1}^{\infty}\frac{(2r-1)^{2m-1}(-1)^r q^{r-\frac{1}{2}}}{1+q^{2r-1}}, \qquad (5.144)$$

$$\hat{T}_{2m-2}(q) := \sum_{r=1}^{\infty}\frac{(2r-1)^{2m-2}(-1)^r q^{r-\frac{1}{2}}}{1-q^{2r-1}}. \qquad (5.145)$$

We have the following theorem.

Theorem 5.19. *Let $z := 2K(k)/\pi \equiv 2K/\pi$, as in* (2.1), *with k the modulus. Take $k' := \sqrt{1-k^2}$ and q as in* (2.4). *Let the Bernoulli numbers B_n and Euler numbers E_n be defined by* (2.60) *and* (2.61), *respectively. Take $U_{2m-1}(-q)$, $G_{2m+1}(-q)$, $R_{2m-2}(q)$, $C_{2m-1}(q)$, $D_{2m+1}(q)$, $T_{2m-2}(q)$, $N_{2m}(q)$, and $N_0(q)$ to be the Lambert series in Theorem* 2.4. *Let $\hat{C}_{2m-1}(q)$ and $\hat{T}_{2m-2}(q)$ be the Lambert series in* (5.144) *and* (5.145), *respectively.*

Take $H_n^{(m)}(\{c_\nu\})$ to be the $n \times n$ determinants in Definition 3.3. Let $n = 1, 2, 3, \ldots$. We then have the following expansions.

$$k^{n^2-n} z^{2n^2-n} = \left\{4^{n^2} \prod_{r=1}^{n-1} (2r)!^{-2}\right\} \cdot H_n^{(1)}(\{N_{2\nu-2}(q)\}) \tag{5.146a}$$

$$+ \left\{4^{n^2-1} \prod_{r=1}^{n-1} (2r)!^{-2}\right\} \cdot H_{n-1}^{(3)}(\{N_{2\nu-2}(q)\}), \tag{5.146b}$$

$$(k')^{n^2-n} z^{2n^2-n} = \left\{(-1)^n 4^n \prod_{r=1}^{n-1} (2r)!^{-2}\right\} \cdot H_n^{(1)}\left(\left\{R_{2\nu-2}(-q) - (-1)^{\nu-1} \cdot \tfrac{1}{4} \cdot |E_{2\nu-2}|\right\}\right), \tag{5.147}$$

$$k^{n^2} z^{2n^2-n} = \left\{4^n \prod_{r=1}^{n-1} (2r)!^{-2}\right\} \cdot H_n^{(1)}(\{T_{2\nu-2}(q)\}), \tag{5.148}$$

$$k^{n^2} (k')^{n^2-n} z^{2n^2-n} = \left\{(-1)^{n(n+1)/2} 4^n \prod_{r=1}^{n-1} (2r)!^{-2}\right\} \cdot H_n^{(1)}(\{\hat{T}_{2\nu-2}(q)\}), \tag{5.149}$$

$$(k')^{n^2} z^{2n^2-n} = \left\{(-1)^n 4^n \prod_{r=1}^{n-1} (2r)!^{-2}\right\} \cdot H_n^{(1)}\left(\left\{R_{2\nu-2}(q) - (-1)^{\nu-1} \cdot \tfrac{1}{4} \cdot |E_{2\nu-2}|\right\}\right), \tag{5.150}$$

$$k^{n^2-n} (k')^{n^2} z^{2n^2-n} = \left\{(-1)^{n(n-1)/2} 4^{n^2} \prod_{r=1}^{n-1} (2r)!^{-2}\right\} \cdot H_n^{(1)}(\{N_{2\nu-2}(-q)\}) \tag{5.151a}$$

$$+ \left\{(-1)^{n(n-1)/2} 4^{n^2-1} \prod_{r=1}^{n-1} (2r)!^{-2}\right\} \cdot H_{n-1}^{(3)}(\{N_{2\nu-2}(-q)\}), \tag{5.151b}$$

$$(k')^{n^2/2} (1 + k')^{n^2-n} z^{2n^2-n} = \left\{(-1)^n 2^{n(n+1)} \prod_{r=1}^{n-1} (2r)!^{-2}\right\} \cdot H_n^{(1)}\left(\left\{R_{2\nu-2}(q^2) - (-1)^{\nu-1} \cdot \tfrac{1}{4} \cdot |E_{2\nu-2}|\right\}\right), \tag{5.152}$$

$$z^{2n^2} = \left\{(-1)^n 2^{2n^2+n} \prod_{r=1}^{2n-1} (r!)^{-1}\right\} \cdot H_n^{(1)}\left(\left\{U_{2\nu-1}(-q) - (-1)^{\nu-1} \frac{(2^{2\nu}-1)}{4\nu} \cdot |B_{2\nu}|\right\}\right), \tag{5.153}$$

$$(kz)^{2n^2} = \left\{4^{n(n+1)} \prod_{r=1}^{2n-1} (r!)^{-1}\right\} \cdot H_n^{(1)}(\{C_{2\nu-1}(q)\}), \tag{5.154}$$

$$(k'z)^{2n^2} = \left\{(-1)^n 2^{2n^2+n} \prod_{r=1}^{2n-1} (r!)^{-1}\right\}$$

$$\cdot H_n^{(1)}\left(\left\{U_{2\nu-1}(q) - (-1)^{\nu-1}\frac{(2^{2\nu}-1)}{4\nu}\cdot|B_{2\nu}|\right\}\right), \quad (5.155)$$

$$(kk')^{n^2} z^{2n^2} = \left\{(-1)^{n(n+1)/2} 4^n \prod_{r=1}^{2n-1} (r!)^{-1}\right\} \cdot H_n^{(1)}(\{\hat{C}_{2\nu-1}(q)\}), \quad (5.156)$$

$$(k')^{n^2} z^{2n^2} = \left\{(-1)^n 2^{2n^2+n} \prod_{r=1}^{2n-1} (r!)^{-1}\right\}$$

$$\cdot H_n^{(1)}\left(\left\{U_{2\nu-1}(q^2) - (-1)^{\nu-1}\frac{(2^{2\nu}-1)}{4\nu}\cdot|B_{2\nu}|\right\}\right), \quad (5.157)$$

$$k^{n^2} z^{2n^2} = \left\{4^n \prod_{r=1}^{2n-1} (r!)^{-1}\right\} \cdot H_n^{(1)}(\{C_{2\nu-1}(\sqrt{q})\}), \quad (5.158)$$

$$k^{n^2} z^{2n^2+n} = \left\{4^n \prod_{r=1}^{n} (2r-1)!^{-2}\right\} \cdot H_n^{(1)}(\{T_{2\nu}(q)\}), \quad (5.159)$$

$$(k')^{n^2} z^{2n^2+n} = \left\{4^n \prod_{r=1}^{n} (2r-1)!^{-2}\right\}$$

$$\cdot H_n^{(1)}\left(\left\{R_{2\nu}(q) - (-1)^{\nu}\cdot\tfrac{1}{4}\cdot|E_{2\nu}|\right\}\right), \quad (5.160)$$

$$k^{n^2+n} z^{2n^2+n} = \left\{4^{n(n+1)} \prod_{r=1}^{n} (2r-1)!^{-2}\right\} \cdot H_n^{(1)}(\{N_{2\nu}(q)\}), \quad (5.161)$$

$$k^{n^2}(k')^{n^2+n} z^{2n^2+n} = \left\{(-1)^{n(n+1)/2} 4^n \prod_{r=1}^{n} (2r-1)!^{-2}\right\} \cdot H_n^{(1)}(\{\hat{T}_{2\nu}(q)\}), \quad (5.162)$$

$$(k')^{n^2+n} z^{2n^2+n} = \left\{4^n \prod_{r=1}^{n} (2r-1)!^{-2}\right\}$$

$$\cdot H_n^{(1)}\left(\left\{R_{2\nu}(-q) - (-1)^{\nu}\cdot\tfrac{1}{4}\cdot|E_{2\nu}|\right\}\right), \quad (5.163)$$

$$k^{n^2+n}(k')^{n^2} z^{2n^2+n} = \left\{(-1)^{n(n+1)/2} 4^{n(n+1)} \prod_{r=1}^{n} (2r-1)!^{-2}\right\}$$

$$\cdot H_n^{(1)}(\{N_{2\nu}(-q)\}), \quad (5.164)$$

$$k^{n^2+n} z^{2n^2+2n} = \left\{2^{2n^2+3n} \prod_{r=1}^{2n} (r!)^{-1}\right\} \cdot H_n^{(1)}(\{D_{2\nu+1}(q)\}), \quad (5.165)$$

$$k^{n^2+n}(k')^{n^2+n}z^{2n^2+2n} = \left\{(-1)^{n(n+1)/2}2^{2n^2+3n}\prod_{r=1}^{2n}(r!)^{-1}\right\} \cdot H_n^{(1)}(\{D_{2\nu+1}(-q)\}), \tag{5.166}$$

$$(kz)^{2n^2+2n} = \left\{2^{4n^2+5n}\prod_{r=1}^{2n}(r!)^{-1}\right\}\cdot H_n^{(1)}(\{D_{2\nu+1}(q^2)\}), \tag{5.167}$$

$$(k'z)^{2n^2+2n} = \left\{2^{2n^2+3n}\prod_{r=1}^{2n}(r!)^{-1}\right\} \cdot H_n^{(1)}\left(\left\{G_{2\nu+1}(q)-(-1)^{\nu}\frac{(2^{2\nu+2}-1)}{4(\nu+1)}\cdot|B_{2\nu+2}|\right\}\right), \tag{5.168}$$

$$z^{2n^2+2n} = \left\{2^{2n^2+3n}\prod_{r=1}^{2n}(r!)^{-1}\right\}\cdot H_n^{(1)}\left(\left\{G_{2\nu+1}(-q) - (-1)^{\nu}\frac{(2^{2\nu+2}-1)}{4(\nu+1)}\cdot|B_{2\nu+2}|\right\}\right). \tag{5.169}$$

Proof: Ten of the identities in (5.146)–(5.169) are immediate from the above analysis in this section. In particular, (5.146), (5.148), (5.150), (5.153), (5.154), (5.159), (5.160), (5.161), (5.165), (5.169) are just (5.129) and (5.131), (5.111) and (5.113), (5.58) and (5.60), (5.24) and (5.26), (5.97) and (5.99), (5.119) and (5.121), (5.78) and (5.80), (5.137) and (5.139), (5.101) and (5.103), (5.41) and (5.43), respectively.

Equation (5.158) results from applying the Gauß transformation ($q \mapsto \sqrt{q}, kz \mapsto 2\sqrt{k}z$) to (5.154).

Equations (5.147), (5.149), (5.151), (5.155), (5.156), (5.162), (5.163), (5.164), (5.166), (5.168) follow by applying Jacobi's transformation ($q \mapsto -q, k'z \mapsto z, z \mapsto k'z, kz \mapsto ikz$) to Eqs. (5.150), (5.148), (5.146), (5.153), (5.158), (5.159), (5.160), (5.161), (5.165), (5.169), respectively.

Finally, Eqs. (5.152), (5.157), (5.167) follow by applying Landen's transformation ($q \mapsto q^2, k'z \mapsto \sqrt{k'}z, z \mapsto \frac{1+k'}{2}z, kz \mapsto \frac{1-k'}{2}z$) to Eqs. (5.150), (5.155), and (5.165), respectively. □

We have listed the identities in Theorem 5.19 in the order in which their $n = 1$ cases appear in [117]. Both (5.146) and (5.147), (5.148) and (5.149), (5.150) and (5.151), generalize, respectively, [117, Eq. (4) (see also (30)), Eq. (5) (see also (31)), Eq. (6) (see also (32)), Section 40]. Equations (5.152)–(5.169) generalize (in order) the following Eqs. in [117]: (7)–(13) (see also (33)–(39)), (40)–(45), of Section 40; (3)–(5), of Section 41; (7), (8) of Section 42.

The power $4n(n+1)$ in (5.37), and powers $2n^2+n$ and $2n^2-n$ of z in Theorem 5.19 also occur as the powers of $\eta(q)$ in Macdonald's expansions corresponding to A_ℓ ($\ell \geq 1$) (take $\ell = 2n$ here), B_ℓ ($\ell \geq 3$), C_ℓ ($\ell \geq 2$), BC_ℓ ($\ell \geq 1$), and D_ℓ ($\ell \geq 4$). These expansions with powers $\ell^2+2\ell, 2\ell^2+\ell, 2\ell^2+\ell, 2\ell^2-\ell$, and $2\ell^2-\ell$, can be found in pages 134–135

(Eq. (6)(a)–(b)), 135 (Eq. (6)(a)), 136 (Eq. (6)), 137–138 (Eq. (6)(c)), and 138 (Eq. (6)), respectively, of [151].

Just after Eq. (45) in Section 40 of [117], Jacobi described how to immediately obtain Lambert series expansions for z^3, $(kz)^3$, and $(k'z)^3$. The coefficient of q^N in these expansions quickly leads to elegant formulas for counting the number of ways that N can be represented as a sum of 6 squares, and also as a sum of 6 triangular numbers. Later, Smith obtained both the analytic and combinatorial formulas for 2, 4, 6, 8 squares and triangles in [213, Section 127, pp. 306–311]. See also [214, p. 206] for a discussion of the 6 and 8 squares work of Jacobi, Eisenstein, and M. Liouville. Glaisher derived the elliptic function and combinatorial formulations of the 6 squares and 6 triangles results in [86, pp. 9–10]. More recent derivations of the 6 squares formulas can be found in Ramanujan [194, Eqs. (135), (136), (145)–(147), Table VI. (entry 1), pp. 158–159], K. Ananda-Rau [3, p. 86], Carlitz [37], Grosswald [94, Eq. (9.19), p. 121], Hardy and Wright [102, pp. 314–315], Kac and Wakimoto [120, Eq. (0.3), p. 416; Example 5.2, p. 444], and Nathanson [182, p. 424; Section 14.5, pp. 436–440]. Several more recent derivations of the 6 triangles formulas appear in Ramanujan [195, Eq. (3.23) (Lambert series version), p. 356], Kac and Wakimoto [120, Example 5.2, p. 444], Ono et al. [184, Theorem 4, p. 81], Berndt [24, Entry 6 (Lambert series version); Corollary 6.2 (arithmetical formula)], Andrews and Berndt [7], and Huard et al. [108, Theorem 11, Section 6].

Motivated by Jacobi and Glaisher's treatment of the 6 squares and triangles formulas we next set $n = 2$ in Eqs. (5.146)–(5.151) of Theorem 5.19 and then derive elegant new Lambert series expansions for z^6, k^2z^6, k^4z^6, and $(kz)^6$, which do not use cusp forms. We have the following theorem.

Theorem 5.20. *Let $z := 2\mathbf{K}(k)/\pi \equiv 2\mathbf{K}/\pi$, as in* (2.1), *with k the modulus. Take q as in* (2.4). *Let $T_{2m-2}(q)$, $\hat{T}_{2m-2}(q)$, $N_{2m-2}(q)$, and $R_{2m-2}(-q)$ be the Lambert series determined by* (2.85), (5.145), (2.86) *and* (2.87), *and* (2.82), *respectively. We then have the following expansions.*

$$(kz)^6 = 4\det\begin{vmatrix} T_0(q) & T_2(q) \\ T_2(q) & T_4(q) \end{vmatrix} + 4\det\begin{vmatrix} \hat{T}_0(q) & \hat{T}_2(q) \\ \hat{T}_2(q) & \hat{T}_4(q) \end{vmatrix}, \tag{5.170}$$

$$k^4z^6 = 4\det\begin{vmatrix} T_0(q) & T_2(q) \\ T_2(q) & T_4(q) \end{vmatrix}, \tag{5.171}$$

$$k^2z^6 = 4^2N_4(q) + 4^3\det\begin{vmatrix} N_0(q) & N_2(q) \\ N_2(q) & N_4(q) \end{vmatrix} \tag{5.172}$$

$$z^6 = 4^2N_4(q) + 4^3\det\begin{vmatrix} N_0(q) & N_2(q) \\ N_2(q) & N_4(q) \end{vmatrix} + \tfrac{1}{4}\det\begin{vmatrix} 4R_0(-q)-1 & 4R_2(-q)+1 \\ 4R_2(-q)+1 & 4R_4(-q)-5 \end{vmatrix} \tag{5.173}$$

Proof: Equations (5.172) and (5.171) are the $n = 2$ cases of (5.146) and (5.148), respectively. Equation (5.170) is immediate from subtracting the $n = 2$ case of (5.149) from (5.171). Equation (5.173) results from adding the $n = 2$ case of (5.147) to (5.172). □

Equations (5.170) and (5.173) can be viewed as additional two-dimensional generalizations of Jacobi's 2 triangles and 2 squares results in [117, Eqs. (5) and (4), Section 40]. Furthermore, (5.170) and (5.173) are not the same as the classical results of Liouvlle [144], [150, (setting for number theory publications), pp. 227–230; (list of number theory publications), pp. 805–812], [56, p. 306], and Glaisher [86, p. 36], [87, pp. 201–202], or the more recent treatments of Ramanujan [194, Eqs. (135), (136), (145)–(147), Table VI. (entry 2), pp. 158–159], Lomadze [147, Theorem 6, p. 264], Gundlach [95, p. 196], Grosswald [94, Eq. (9.19), p. 121], Kac and Wakimoto [120, $m = 2$ case of Example 5.3, p. 444], and Ono et al. [184, Theorem 7, p. 85]. Note that Theorem 7 of [184, p. 85] and its resulting congruence mod 256 also appears at the very end of Section 61 of Glaisher's article [86, p. 37].

In order to generalize the above classical work of Jacobi, Glaisher, and Ramanujan from 6 squares and 6 triangles to 20 squares and 20 triangles we first set $n = 2$ in the 6 Eqs. (5.159)–(5.164) of Theorem 5.19 and then derive the Lambert series expansions for z^{10}, k^2z^{10}, k^4z^{10}, k^6z^{10}, k^8z^{10}, and $(kz)^{10}$ in the following theorem.

Theorem 5.21. *Let $z := 2K(k)/\pi \equiv 2K/\pi$, as in (2.1), with k the modulus. Take q as in (2.4). Let $T_{2m-2}(q)$, $\hat{T}_{2m-2}(q)$, $N_{2m-2}(q)$, and $R_{2m-2}(q)$ be the Lambert series determined by (2.85), (5.145), (2.86) and (2.87), and (2.82), respectively. We then have the following expansions.*

$$(kz)^{10} = \frac{2^{10}}{3}\left\{\det\begin{vmatrix} N_2(q) & N_4(q) \\ N_4(q) & N_6(q)\end{vmatrix} - \det\begin{vmatrix} N_2(-q) & N_4(-q) \\ N_4(-q) & N_6(-q)\end{vmatrix}\right\}$$

$$- \frac{2^3}{3^2}\left\{\det\begin{vmatrix} T_2(q) & T_4(q) \\ T_4(q) & T_6(q)\end{vmatrix} + \det\begin{vmatrix} \hat{T}_2(q) & \hat{T}_4(q) \\ \hat{T}_4(q) & \hat{T}_6(q)\end{vmatrix}\right\}, \tag{5.174}$$

$$k^8z^{10} = \frac{2^{10}}{3^2}\left\{2\det\begin{vmatrix} N_2(q) & N_4(q) \\ N_4(q) & N_6(q)\end{vmatrix} - \det\begin{vmatrix} N_2(-q) & N_4(-q) \\ N_4(-q) & N_6(-q)\end{vmatrix}\right\}$$

$$- \frac{2^2}{3^2}\left\{\det\begin{vmatrix} T_2(q) & T_4(q) \\ T_4(q) & T_6(q)\end{vmatrix} + \det\begin{vmatrix} \hat{T}_2(q) & \hat{T}_4(q) \\ \hat{T}_4(q) & \hat{T}_6(q)\end{vmatrix}\right\} \tag{5.175}$$

$$k^6z^{10} = \frac{2^{10}}{3^2}\det\begin{vmatrix} N_2(q) & N_4(q) \\ N_4(q) & N_6(q)\end{vmatrix}, \tag{5.176}$$

$$k^4z^{10} = \frac{2^2}{3^2}\det\begin{vmatrix} T_2(q) & T_4(q) \\ T_4(q) & T_6(q)\end{vmatrix}, \tag{5.177}$$

$$k^2 z^{10} = \frac{2^3}{3^2}\det\begin{vmatrix} T_2(q) & T_4(q) \\ T_4(q) & T_6(q) \end{vmatrix} - \frac{2^{10}}{3^2}\det\begin{vmatrix} N_2(q) & N_4(q) \\ N_4(q) & N_6(q) \end{vmatrix}$$
$$+ \frac{1}{6^2}\left\{\det\begin{vmatrix} 4R_2(q)+1 & 4R_4(q)-5 \\ 4R_4(q)-5 & 4R_6(q)+61 \end{vmatrix}\right.$$
$$\left. - \det\begin{vmatrix} 4R_2(-q)+1 & 4R_4(-q)-5 \\ 4R_4(-q)-5 & 4R_6(-q)+61 \end{vmatrix}\right\}, \tag{5.178}$$

$$z^{10} = \frac{2^2}{3}\det\begin{vmatrix} T_2(q) & T_4(q) \\ T_4(q) & T_6(q) \end{vmatrix} - \frac{2^{11}}{3^2}\det\begin{vmatrix} N_2(q) & N_4(q) \\ N_4(q) & N_6(q) \end{vmatrix}$$
$$+ \frac{1}{18}\left\{\frac{3}{2}\det\begin{vmatrix} 4R_2(q)+1 & 4R_4(q)-5 \\ 4R_4(q)-5 & 4R_6(q)+61 \end{vmatrix}\right.$$
$$\left. - \det\begin{vmatrix} 4R_2(-q)+1 & 4R_4(-q)-5 \\ 4R_4(-q)-5 & 4R_6(-q)+61 \end{vmatrix}\right\}. \tag{5.179}$$

Proof: Let $\alpha_1, \alpha_2, \alpha_3, \alpha_4, \alpha_5, \alpha_6$ denote the $n = 2$ cases, respectively, of Eqs. (5.159), (5.160), (5.161), (5.162), (5.163), and (5.164). Then, Eqs. (5.176) and (5.177) are α_3 and α_1, respectively. Equation (5.175) is $(-\alpha_1 + 2\alpha_3 + \alpha_4 + \alpha_6)$. Equation (5.174) results from $(-2\alpha_1 + 2\alpha_4 + 3\alpha_3 + 3\alpha_6)$. Equation (5.178) follows from $(2\alpha_1 + \alpha_2 - \alpha_3 - \alpha_5)$. Finally, Eq. (5.179) is a consequence of $(3\alpha_1 + 3\alpha_2 - 2\alpha_3 - 2\alpha_5)$. □

The formula for z^{10} in (5.179) is not the same as the classical formulas of Ramanujan [194, Eqs. (135), (136), (145)–(147), Table VI. (entry 5), pp. 158–159], Rankin [202], and Lomadze [147, Theorem 7, p. 266]. Furthermore, it appears likely that the Schur function expansion in Theorem 6.7 applied to the determinants on the right-hand-side of (5.174) will lead to a proof of the $m = 2$ case of Conjecture 5.1 in [120, p. 445]. The equivalence should be nontrivial.

We have developed many more consequences of Theorem 5.19. This work will appear elsewhere.

We next survey our χ_n determinant identities. The derivation of several of these identities requires the simplification

$$\sum_{r=1}^{\infty} \frac{(2r-1)q^{2r-1}}{1-q^{2r-1}} = \sum_{r=1}^{\infty} \frac{rq^r}{1+q^r}. \tag{5.180}$$

Equation (5.180) is the $x \mapsto \sqrt{q}$ case of the identity at the top of page 34 of [94]. Apply the partial fraction $q^r/(1+q^r) = q^r/(1-q^r) - 2q^{2r}/(1-q^{2r})$ termwise to the right-hand-side, and then cancell the even index terms. This simplification came up in the context of some Lambert series identities of Ramanujan.

We also need the observation that the proof of Lemma 5.1 immediately implies that Eq. (5.3) remains valid if Hankel determinants are replaced by χ_n determinants. That is, we have the following lemma.

Lemma 5.22. *Let $v_1, \ldots, v_{2n}$ and $w_1, \ldots, w_{2n}$ be indeterminate, and let $n = 1, 2, 3, \ldots$. Suppose that $\chi_n(\{w_\nu\})$ and $\chi_n(\{v_\nu\})$ are the determinants of the $n \times n$ square matrices given by* (3.56)*, with the $n = 1$ cases equal to w_2 and v_2, respectively. Let $M_{n,S}$ be the $n \times n$ matrix whose i-th row is*

$$v_i + w_i, v_{i+1} + w_{i+1}, \ldots, v_{i+n-2} + w_{i+n-2}, v_{i+n} + w_{i+n}, \quad \text{if } i \in S,$$

and

$$v_i, v_{i+1}, \ldots, v_{i+n-2}, v_{i+n}, \quad \text{if } i \notin S. \tag{5.181}$$

Then

$$\chi_n(\{w_\nu\}) = \sum_{\emptyset \subseteq S \subseteq I_n} (-1)^{n-\|S\|} \det(M_{n,S}) \tag{5.182}$$

$$= (-1)^n \chi_n(\{v_\nu\}) + \sum_{p=1}^{n} (-1)^{n-p} \sum_{\substack{\emptyset \subset S \subseteq I_n \\ \|S\|=p}} \det(M_{n,S}). \tag{5.183}$$

We first have the following theorem.

Theorem 5.23. *Let $\vartheta_3(0, -q)$ be determined by* (1.1)*, and let $n = 1, 2, 3, \ldots$. We then have*

$$\vartheta_3(0, -q)^{4n^2} \left[1 + 24 \sum_{r=1}^{\infty} \frac{rq^r}{1+q^r}\right] = \left\{(-1)^{n-1} 2^{2n^2+n+1} \frac{3}{n(4n^2-1)} \prod_{r=1}^{2n-1} (r!)^{-1}\right\} \cdot \chi_n(\{g_\nu\}), \tag{5.184}$$

where $\chi_n(\{g_\nu\})$ is the determinant of the $n \times n$ matrix whose i-th row is

$$g_i, g_{i+1}, \ldots, g_{i+n-2}, g_{i+n}, \quad \text{for } i = 1, 2, \ldots, n, \tag{5.185}$$

where

$$g_i := U_{2i-1} - c_i, \tag{5.186}$$

when $n \geq 2$. If $n = 1$, then $\chi_1(\{g_\nu\})$ equals $g_2 = U_3 - c_2$. The U_{2i-1} and c_i are defined by (5.22) *and* (5.23)*, respectively, with B_{2i} the Bernoulli numbers defined by* (2.60).

Proof: Our analysis deals with sc(u, k) dn(u, k). Starting with (2.80), applying row operations and (3.68), appealing to (4.52), and then utilizing both (5.9) and (5.15) we have the following computation.

$$\chi_n\left(\left\{U_{2\nu-1}(-q) - (-1)^{\nu-1}\frac{(2^{2\nu}-1)}{4\nu}\cdot|B_{2\nu}|\right\}\right) \tag{5.187}$$

$$= \chi_n\left(\left\{(-1)^{\nu}\frac{z^{2\nu}}{2^{2\nu+1}}\cdot(sd/c)_\nu(k^2)\right\}\right) \tag{5.188}$$

$$= \left\{(-1)^{n-1}2^{-(2n^2+n+1)}\frac{n(4n^2-1)}{3}(1-2k^2)\prod_{r=1}^{2n-1} r!\right\}\cdot z^{2n^2+2} \tag{5.189}$$

$$= \left\{(-1)^{n-1}2^{-(2n^2+n+1)}\frac{n(4n^2-1)}{3}\prod_{r=1}^{2n-1} r!\right\}$$

$$\cdot\,\vartheta_3(0,q)^{4n^2}\left[1 - 24\sum_{r=1}^{\infty}\frac{(2r-1)q^{2r-1}}{1+q^{2r-1}}\right]. \tag{5.190}$$

Replacing q by $-q$ in (5.187) and (5.190) and then using (5.180) gives

$$\chi_n\left(\left\{U_{2\nu-1}(q) - (-1)^{\nu-1}\frac{(2^{2\nu}-1)}{4\nu}\cdot|B_{2\nu}|\right\}\right) \tag{5.191a}$$

$$= \left\{(-1)^{n-1}2^{-(2n^2+n+1)}\frac{n(4n^2-1)}{3}\prod_{r=1}^{2n-1} r!\right\} \tag{5.191b}$$

$$\cdot\,\vartheta_3(0,-q)^{4n^2}\left[1 + 24\sum_{r=1}^{\infty}\frac{rq^r}{1+q^r}\right]. \tag{5.191c}$$

Solving for (5.191c) in (5.191) yields (5.184). □

Lemma 5.22 and Theorem 5.23 lead to the χ_n determinant sum identity in the following theorem.

Theorem 5.24. *Let* $n = 1, 2, 3, \ldots$. *Then*

$$\vartheta_3(0,-q)^{4n^2}\left[1 + 24\sum_{r=1}^{\infty}\frac{rq^r}{1+q^r}\right]$$

$$= 1 + \sum_{p=1}^{n}(-1)^{p-1}2^{2n^2+n+1}\frac{3}{n(4n^2-1)}\prod_{r=1}^{2n-1}(r!)^{-1}\sum_{\substack{\emptyset\subset S\subseteq I_n\\ \|S\|=p}}\det(M_{n,S}), \tag{5.192}$$

where $\vartheta_3(0,-q)$ *is determined by* (1.1), *and* $M_{n,S}$ *is the* $n\times n$ *matrix whose* i*-th row is*

$$\begin{array}{ll} U_{2i-1}, U_{2(i+1)-1}, \ldots, U_{2(i+n-2)-1}, U_{2(i+n)-1}, & \text{if } i\in S,\\ c_i, c_{i+1}, \ldots, c_{i+n-2}, c_{i+n}, & \text{if } i\notin S, \end{array} \tag{5.193}$$

when $n \geq 2$. If $n = 1$, then $M_{n,S}$ is the 1×1 matrix

$$(U_3), \quad \textit{since } S = \{1\} \textit{ and } 1 \in S. \tag{5.194}$$

The U_{2i-1} and c_i are defined by (5.22) *and* (5.23), *respectively, with B_{2i} the Bernoulli numbers defined by* (2.60).

Proof: Specialize v_i and w_i in Eq. (5.183) of Lemma 5.22 as in (5.31) and (5.32), respectively. Recall (5.33).

Equating (5.188) and (5.190) it is immediate that

$$\chi_n(\{w_\nu\}) = \left\{(-1)^{n-1} 2^{-(2n^2+n+1)\frac{n(4n^2-1)}{3}} \prod_{r=1}^{2n-1} r!\right\} \cdot \vartheta_3(0,q)^{4n^2}\left[1 - 24\sum_{r=1}^{\infty}\frac{(2r-1)q^{2r-1}}{1+q^{2r-1}}\right]. \tag{5.195}$$

From (4.60) we have

$$\chi_n(\{v_\nu\}) = -^{\frac{n(4n^2-1)}{3}} 2^{-(2n^2+n+1)} \prod_{r=1}^{2n-1} r!. \tag{5.196}$$

Equation (5.192) is now a direct consequence of applying the determinant evaluations in (5.195) and (5.196), replacing q by $-q$, using (5.180), and then multiplying both sides of the resulting transformation of (5.183) by

$$(-1)^{n-1} 2^{2n^2+n+1} \frac{3}{n(4n^2-1)} \prod_{r=1}^{2n-1} (r!)^{-1}, \tag{5.197}$$

and simplifying. □

The process of "obtaining a formula by duplication" from [21, p. 125] applied to the $n = 1$ case of Theorem 5.24 gives the identity in [259, Eq. (4) of Table 1(x), p. 201]. The $n = 2$ case of Theorem 5.24 immediately leads to

$$\vartheta_3(0,-q)^{16}\left[1 + 24\sum_{r=1}^{\infty}\frac{rq^r}{1+q^r}\right] = 1 - \tfrac{8}{15}[17U_1 + 4U_3 - 2U_5 - 4U_7] - \tfrac{256}{15}[U_1U_7 - U_3U_5], \tag{5.198}$$

where U_{2i-1} is defined by (5.22), and $\vartheta_3(0,-q)$ is determined by (1.1).

We next have the following theorem.

Theorem 5.25. *Let $\vartheta_3(0,-q)$ be determined by* (1.1), *and let $n = 1, 2, 3, \ldots$. We then have*

$$\vartheta_3(0,-q)^{4n(n+1)}\left[1+24\sum_{r=1}^{\infty}\frac{rq^r}{1+q^r}\right]$$

$$= \left\{(-1)2^{2n^2+3n}\frac{3}{n(n+1)(2n+1)}\prod_{r=1}^{2n}(r!)^{-1}\right\}\cdot\chi_n(\{g_\nu\}), \tag{5.199}$$

where $\chi_n(\{g_\nu\})$ is the determinant of the $n\times n$ matrix whose i-th row is

$$g_i, g_{i+1}, \ldots, g_{i+n-2}, g_{i+n}, \quad \textit{for } i = 1, 2, \ldots, n, \tag{5.200}$$

where

$$g_i := G_{2i+1} - a_i, \tag{5.201}$$

when $n \geq 2$. If $n = 1$, then $\chi_1(\{g_\nu\})$ equals $g_2 = G_5 - a_2$. The G_{2i+1} and a_i are defined by (5.39) *and* (5.40), *respectively, with B_{2i+2} the Bernoulli numbers defined by* (2.60).

Proof: Our analysis deals with $\mathrm{sc}^2(u,k)\,\mathrm{dn}^2(u,k)$. Starting with (2.81), applying row operations and (3.68), appealing to (4.54), and then utilizing both (5.9) and (5.15) we have the following computation.

$$\chi_n\left(\left\{G_{2\nu+1}(-q) - (-1)^\nu\frac{(2^{2\nu+2}-1)}{4(\nu+1)}\cdot|B_{2\nu+2}|\right\}\right) \tag{5.202}$$

$$= \chi_n\left(\left\{(-1)^{\nu-1}\frac{z^{2\nu+2}}{2^{2\nu+3}}\cdot(s^2d^2/c^2)_\nu(k^2)\right\}\right) \tag{5.203}$$

$$= \left\{-2^{-(2n^2+3n)}\frac{n(n+1)(2n+1)}{3}(1-2k^2)\prod_{r=1}^{2n}r!\right\}\cdot z^{2n^2+2n+2} \tag{5.204}$$

$$= \left\{-2^{-(2n^2+3n)}\frac{n(n+1)(2n+1)}{3}\prod_{r=1}^{2n}r!\right\}$$

$$\cdot\,\vartheta_3(0,q)^{4n(n+1)}\left[1-24\sum_{r=1}^{\infty}\frac{(2r-1)q^{2r-1}}{1+q^{2r-1}}\right]. \tag{5.205}$$

Replacing q by $-q$ in (5.202) and (5.205) and then using (5.180) gives

$$\chi_n\left(\left\{G_{2\nu+1}(q) - (-1)^\nu\frac{(2^{2\nu+2}-1)}{4(\nu+1)}\cdot|B_{2\nu+2}|\right\}\right) \tag{5.206a}$$

$$= \left\{-2^{-(2n^2+3n)}\frac{n(n+1)(2n+1)}{3}\prod_{r=1}^{2n}r!\right\} \tag{5.206b}$$

$$\cdot\,\vartheta_3(0,-q)^{4n(n+1)}\left[1+24\sum_{r=1}^{\infty}\frac{rq^r}{1+q^r}\right]. \tag{5.206c}$$

Solving for (5.206c) in (5.206) yields (5.199). □

Lemma 5.22 and Theorem 5.25 lead to the χ_n determinant sum identity in the following theorem.

Theorem 5.26. *Let $n = 1, 2, 3, \ldots$. Then*

$$\vartheta_3(0,-q)^{4n(n+1)}\left[1+24\sum_{r=1}^{\infty}\frac{rq^r}{1+q^r}\right]$$

$$= 1+\sum_{p=1}^{n}(-1)^{n-p+1}2^{2n^2+3n}\frac{3}{n(n+1)(2n+1)}\prod_{r=1}^{2n}(r!)^{-1}\sum_{\substack{\emptyset\subset S\subseteq I_n\\ \|S\|=p}}\det(M_{n,S}), \quad (5.207)$$

where $\vartheta_3(0,-q)$ is determined by (1.1), *and $M_{n,S}$ is the $n\times n$ matrix whose i-th row is*

$$\begin{aligned}&G_{2i+1}, G_{2(i+1)+1}, \ldots, G_{2(i+n-2)+1}, G_{2(i+n)+1}, \quad \text{if } i\in S,\\ &a_i, a_{i+1}, \ldots, a_{i+n-2}, a_{i+n}, \quad \text{if } i\notin S,\end{aligned} \quad (5.208)$$

when $n\geq 2$. If $n=1$, then $M_{n,S}$ is the 1×1 matrix

$$(G_5), \quad \text{since} \quad S=\{1\} \quad \text{and} \quad 1\in S. \quad (5.209)$$

The G_{2i+1} and a_i are defined by (5.39) *and* (5.40), *respectively, with B_{2i+2} the Bernoulli numbers defined by* (2.60).

Proof: Specialize v_i and w_i in Eq. (5.183) of Lemma 5.22 as in (5.48) and (5.49), respectively. Recall (5.50).

Equating (5.203) and (5.205) it is immediate that

$$\chi_n(\{w_\nu\}) = \left\{-2^{-(2n^2+3n)}\frac{n(n+1)(2n+1)}{3}\prod_{r=1}^{2n}r!\right\}$$

$$\cdot\,\vartheta_3(0,q)^{4n(n+1)}\left[1-24\sum_{r=1}^{\infty}\frac{(2r-1)q^{2r-1}}{1+q^{2r-1}}\right]. \quad (5.210)$$

From (4.61) we have

$$\chi_n(\{v_\nu\}) = (-1)^{n-1}\cdot\frac{n(n+1)(2n+1)}{3}2^{-(2n^2+3n)}\prod_{r=1}^{2n}r!. \quad (5.211)$$

Equation (5.207) is now a direct consequence of applying the determinant evaluations in (5.210) and (5.211), replacing q by $-q$, using (5.180), and then multiplying both sides of the resulting transformation of (5.183) by

$$-2^{2n^2+3n}\frac{3}{n(n+1)(2n+1)}\prod_{r=1}^{2n}(r!)^{-1}, \quad (5.212)$$

and simplifying. □

The process of "obtaining a formula by duplication" from [21, p. 125] applied to the $n = 1$ case of Theorem 5.26 gives the identity in [259, Eq. (5) of Table 1(ii), p. 197]. The $n = 2$ case of Theorem 5.26 immediately leads to

$$\vartheta_3(0,-q)^{24}\left[1+24\sum_{r=1}^{\infty}\frac{rq^r}{1+q^r}\right] = 1+\tfrac{8}{45}\left[124G_3+17G_5-4G_7-2G_9\right] - \tfrac{256}{45}\left[G_3G_9-G_5G_7\right], \tag{5.213}$$

where G_{2i+1} is defined by (5.39), and $\vartheta_3(0,-q)$ is determined by (1.1).

We next have the following theorem.

Theorem 5.27. *Let $\vartheta_3(0,-q)$ be determined by* (1.1), *and let $n = 1, 2, 3, \ldots$. We then have*

$$\vartheta_3(0,q)^{2n(n-1)}\vartheta_3(0,-q)^{2n^2}\cdot\left\{2n\left[2+24\sum_{r=1}^{\infty}\frac{2rq^{2r}}{1+q^{2r}}\right]-\left[1-24\sum_{r=1}^{\infty}\frac{(2r-1)q^{2r-1}}{1+q^{2r-1}}\right]\right\}$$
$$=\left\{(-1)^{n-1}2^{2n}\tfrac{3}{n(2n-1)}\prod_{r=1}^{n-1}(2r)!^{-2}\right\}\cdot\chi_n(\{g_\nu\}), \tag{5.214}$$

where $\chi_n(\{g_\nu\})$ is the determinant of the $n\times n$ matrix whose i-th row is

$$g_i, g_{i+1}, \ldots, g_{i+n-2}, g_{i+n}, \quad \text{for } i = 1, 2, \ldots, n, \tag{5.215}$$

where

$$g_i := R_{2i-2} - b_i, \tag{5.216}$$

when $n \geq 2$. If $n = 1$, then $\chi_1(\{g_\nu\})$ equals $g_2 = R_2 - b_2$. The R_{2i-2} and b_i are defined by (5.56) *and* (5.57), *respectively, with E_{2i-2} the Euler numbers defined by* (2.61).

Proof: Our analysis deals with $\mathrm{nc}(u,k)$. Starting with (2.82), applying row operations and (3.69), appealing to (4.48), and then utilizing (5.9), (5.10), (5.14), and (5.15) we have the following computation.

$$\chi_n\left(\left\{R_{2\nu-2}(q)-(-1)^{\nu-1}\cdot\tfrac{1}{4}\cdot|E_{2\nu-2}|\right\}\right) \tag{5.217}$$
$$=\chi_n\left(\left\{(-1)^{\nu}\frac{z^{2\nu-1}}{4}\sqrt{1-k^2}\cdot(nc)_{\nu-1}(k^2)\right\}\right) \tag{5.218}$$
$$=\left\{(-1)^{n-1}2^{-2n}\tfrac{n(2n-1)}{3}\prod_{r=1}^{n-1}(2r)!^2\right\}\cdot(1-k^2)^{n^2/2}\left[2n(2-k^2)-(1-2k^2)\right]z^{2n^2-n+2} \tag{5.219}$$

$$= \left\{(-1)^{n-1}2^{-2n\frac{n(2n-1)}{3}}\prod_{r=1}^{n-1}(2r)!^2\right\} \tag{5.220a}$$

$$\cdot\,\vartheta_3(0,q)^{2n(n-1)}\vartheta_3(0,-q)^{2n^2} \tag{5.220b}$$

$$\cdot\left\{2n\left[2+24\sum_{r=1}^{\infty}\frac{2rq^{2r}}{1+q^{2r}}\right]-\left[1-24\sum_{r=1}^{\infty}\frac{(2r-1)q^{2r-1}}{1+q^{2r-1}}\right]\right\}. \tag{5.220c}$$

Equating (5.217) with (5.220) and then solving for (5.220b)–(5.220c) yields (5.214). □

Noting that

$$[2n(2-k^2)-(1-2k^2)] = [(2n+1)+2(n-1)(1-k^2)], \tag{5.221}$$

and recalling (5.9) and (5.10), it is sometimes useful to rewrite (5.220c) as

$$[(2n+1)\vartheta_3(0,q)^4+2(n-1)\vartheta_3(0,-q)^4]. \tag{5.222}$$

Lemma 5.22 and Theorem 5.27 lead to the χ_n determinant sum identity in the following theorem.

Theorem 5.28. *Let $n = 1, 2, 3, \ldots$. Then*

$$\vartheta_3(0,q)^{2n(n-1)}\vartheta_3(0,-q)^{2n^2}\cdot\left\{2n\left[2+24\sum_{r=1}^{\infty}\frac{2rq^{2r}}{1+q^{2r}}\right]-\left[1-24\sum_{r=1}^{\infty}\frac{(2r-1)q^{2r-1}}{1+q^{2r-1}}\right]\right\}$$

$$= (4n-1)+\sum_{p=1}^{n}(-1)^{p-1}2^{2n\frac{3}{n(2n-1)}}\prod_{r=1}^{n-1}(2r)!^{-2}\sum_{\substack{\emptyset\subset S\subseteq I_n\\ \|S\|=p}}\det(M_{n,s}), \tag{5.223}$$

where $\vartheta_3(0,-q)$ is determined by (1.1), *and $M_{n,S}$ is the $n\times n$ matrix whose i-th row is*

$$R_{2i-2}, R_{2(i+1)-2}, \ldots, R_{2(i+n-2)-2}, R_{2(i+n)-2}, \quad \textit{if } i\in S,$$

and

$$b_i, b_{i+1}, \ldots, b_{i+n-2}, b_{i+n}, \quad \textit{if } i\notin S, \tag{5.224}$$

when $n\geq 2$. If $n=1$, then $M_{n,S}$ is the 1×1 matrix

$$(R_2), \quad \textit{since } S=\{1\} \textit{ and } 1\in S. \tag{5.225}$$

The R_{2i-2} and b_i are defined by (5.56) *and* (5.57), *respectively, with E_{2i-2} the Euler numbers defined by* (2.61).

Proof: Specialize v_i and w_i in Eq. (5.183) of Lemma 5.22 as in (5.68) and (5.69), respectively. Recall (5.70).

Equating (5.218) and (5.220) it is immediate that

$$\begin{aligned}\chi_n(\{w_\nu\}) = &\left\{(-1)^{n-1}2^{-2n\frac{n(2n-1)}{3}}\prod_{r=1}^{n-1}(2r)!^2\right\}\\ &\cdot \vartheta_3(0,q)^{2n(n-1)}\vartheta_3(0,-q)^{2n^2}\\ &\cdot\left\{2n\left[2+24\sum_{r=1}^{\infty}\frac{2rq^{2r}}{1+q^{2r}}\right]-\left[1-24\sum_{r=1}^{\infty}\frac{(2r-1)q^{2r-1}}{1+q^{2r-1}}\right]\right\}. \qquad (5.226)\end{aligned}$$

From (4.62) we have

$$\chi_n(\{v_\nu\}) = -\frac{n(2n-1)(4n-1)}{3}2^{-2n}\prod_{r=1}^{n-1}(2r)!^2. \qquad (5.227)$$

Equation (5.223) is now a direct consequence of applying the determinant evaluations in (5.226) and (5.227), and then multiplying both sides of the resulting transformation of (5.183) by

$$(-1)^{n-1}2^{2n}\frac{3}{n(2n-1)}\prod_{r=1}^{n-1}(2r)!^{-2}, \qquad (5.228)$$

and simplifying. □

The $n = 1$ case of Theorem 5.28, rewritten using (5.222), gives the identity in [259, Eq. (4) of Table 1(xiv), p. 203]. Similarly, the $n = 2$ case of Theorem 5.28 immediately leads to

$$\begin{aligned}&\vartheta_3(0,q)^4\vartheta_3(0,-q)^8[5\vartheta_3(0,q)^4+2\vartheta_3(0,-q)^4]\\ &\quad = 7-\tfrac{1}{2}[61R_0+5R_2-R_4-R_6]-2[R_0R_6-R_2R_4], \qquad (5.229)\end{aligned}$$

where R_{2i-2} is defined by (5.56), and $\vartheta_3(0,-q)$ is determined by (1.1).

We next have the following theorem.

Theorem 5.29. *Let* $\vartheta_3(0,-q)$ *be determined by* (1.1), *and let* $n = 1, 2, 3, \ldots$. *We then have*

$$\begin{aligned}&\vartheta_3(0,q)^{2n(n+1)}\vartheta_3(0,-q)^{2n^2}\cdot\left\{2n\left[2+24\sum_{r=1}^{\infty}\frac{2rq^{2r}}{1+q^{2r}}\right]+\left[1-24\sum_{r=1}^{\infty}\frac{(2r-1)q^{2r-1}}{1+q^{2r-1}}\right]\right\}\\ &= \left\{-2^{2n}\frac{3}{n(2n+1)}\prod_{r=1}^{n}(2r-1)!^{-2}\right\}\cdot\chi_n(\{g_\nu\}), \qquad (5.230)\end{aligned}$$

where $\chi_n(\{g_\nu\})$ *is the determinant of the* $n \times n$ *matrix whose i-th row is*

$$g_i, g_{i+1}, \ldots, g_{i+n-2}, g_{i+n}, \quad \textit{for } i = 1, 2, \ldots, n, \qquad (5.231)$$

where

$$g_i := R_{2i} - b_{i+1}, \qquad (5.232)$$

when $n \geq 2$. If $n = 1$, then $\chi_1(\{g_v\})$ equals $g_2 = R_4 - b_3$. The R_{2i} and b_{i+1} are determined by (5.56) *and* (5.57), *respectively, with E_{2i} the Euler numbers defined by* (2.61).

Proof: Our analysis deals with sc(u,k) dc(u,k). Starting with (2.90), applying row operations and (3.68), appealing to (4.49), and then utilizing (5.9), (5.10), (5.14), and (5.15) we have the following computation.

$$\chi_n(\{R_{2v}(q) - (-1)^v \cdot \tfrac{1}{4} \cdot |E_{2v}|\}) \tag{5.233}$$

$$= \chi_n\left(\left\{(-1)^{v+1}\frac{z^{2v+1}}{4}\sqrt{1-k^2}\cdot(sd/c^2)_v(k^2)\right\}\right) \tag{5.234}$$

$$= \left\{-2^{-2n}\tfrac{n(2n+1)}{3}\prod_{r=1}^{n}(2r-1)!^2\right\}$$
$$\cdot (1-k^2)^{n^2/2}[2n(2-k^2)+(1-2k^2)]z^{2n^2+n+2} \tag{5.235}$$

$$= \left\{-2^{-2n}\tfrac{n(2n+1)}{3}\prod_{r=1}^{n}(2r-1)!^2\right\} \tag{5.236a}$$

$$\cdot\, \vartheta_3(0,q)^{2n(n+1)}\vartheta_3(0,-q)^{2n^2} \tag{5.236b}$$

$$\cdot\left\{2n\left[2+24\sum_{r=1}^{\infty}\frac{2rq^{2r}}{1+q^{2r}}\right]+\left[1-24\sum_{r=1}^{\infty}\frac{(2r-1)q^{2r-1}}{1+q^{2r-1}}\right]\right\}. \tag{5.236c}$$

Equating (5.233) with (5.236) and then solving for (5.236b)–(5.236c) yields (5.230). □

Noting that

$$[2n(2-k^2)+(1-2k^2)] = [(2n-1)+2(n+1)(1-k^2)], \tag{5.237}$$

and recalling (5.9) and (5.10), it is sometimes useful to rewrite (5.236c) as

$$[(2n-1)\vartheta_3(0,q)^4 + 2(n+1)\vartheta_3(0,-q)^4]. \tag{5.238}$$

Lemma 5.22 and Theorem 5.29 lead to the χ_n determinant sum identity in the following theorem.

Theorem 5.30. *Let $n = 1, 2, 3, \ldots$. Then*

$$\vartheta_3(0,q)^{2n(n+1)}\vartheta_3(0,-q)^{2n^2}\cdot\left\{2n\left[2+24\sum_{r=1}^{\infty}\frac{2rq^{2r}}{1+q^{2r}}\right]+\left[1-24\sum_{r=1}^{\infty}\frac{(2r-1)q^{2r-1}}{1+q^{2r-1}}\right]\right\}$$
$$= (4n+1) + \sum_{p=1}^{n}(-1)^{n-p+1}2^{2n}\tfrac{3}{n(2n+1)}\prod_{r=1}^{n}(2r-1)!^{-2}\sum_{\substack{\emptyset\subset S\subseteq I_n\\ \|S\|=p}}\det(M_{n,S}), \tag{5.239}$$

where $\vartheta_3(0,-q)$ is determined by (1.1)*, and $M_{n,S}$ is the $n \times n$ matrix whose i-th row is*

$$\begin{aligned} &R_{2i}, R_{2(i+1)}, \ldots, R_{2(i+n-2)}, R_{2(i+n)}, \quad \text{if } i \in S, \\ &b_{i+1}, b_{i+2}, \ldots, b_{i+n-1}, b_{i+n+1}, \quad \text{if } i \notin S, \end{aligned} \tag{5.240}$$

when $n \geq 2$. If $n = 1$, then $M_{n,S}$ is the 1×1 matrix

$$(R_4), \quad \text{since } S = \{1\} \text{ and } 1 \in S. \tag{5.241}$$

The R_{2i} and b_{i+1} are determined by (5.56) *and* (5.57)*, respectively, with E_{2i} the Euler numbers defined by* (2.61).

Proof: Specialize v_i and w_i in Eq. (5.183) of Lemma 5.22 as in (5.87) and (5.88), respectively, with the $(nc)_i(k^2)$ in (5.88) replaced by $(sd/c^2)_i(k^2)$. Utilize (2.90) to write $v_i + w_i$ as the Lambert series $R_{2i}(q)$ in (5.89).

Equating (5.234) and (5.236) it is immediate that

$$\begin{aligned} \chi_n(\{w_\nu\}) = &\left\{-2^{-2n\,\frac{n(2n+1)}{3}} \prod_{r=1}^{n} (2r-1)!^2\right\} \\ &\cdot \vartheta_3(0,q)^{2n(n+1)} \vartheta_3(0,-q)^{2n^2} \\ &\cdot \left\{2n\left[2 + 24\sum_{r=1}^{\infty} \frac{2rq^{2r}}{1+q^{2r}}\right] + \left[1 - 24\sum_{r=1}^{\infty} \frac{(2r-1)q^{2r-1}}{1+q^{2r-1}}\right]\right\}. \end{aligned} \tag{5.242}$$

From (4.63) we have

$$\chi_n(\{v_\nu\}) = -(-1)^{n\,\frac{n(2n+1)(4n+1)}{3}} 2^{-2n} \prod_{r=1}^{n} (2r-1)!^2. \tag{5.243}$$

Equation (5.239) is now a direct consequence of applying the determinant evaluations in (5.242) and (5.243), and then multiplying both sides of the resulting transformation of (5.183) by

$$-2^{2n\,\frac{3}{n(2n+1)}} \prod_{r=1}^{n} (2r-1)!^{-2}, \tag{5.244}$$

and simplifying. □

The $n = 1$ case of Theorem 5.30, rewritten using (5.238), gives the identity in [259, Eq. (5) of Table 1(xiv), p. 203]. Similarly, the $n = 2$ case of Theorem 5.30 immediately leads to

$$\begin{aligned} &\vartheta_3(0,q)^{12} \vartheta_3(0,-q)^8 [3\vartheta_3(0,q)^4 + 6\vartheta_3(0,-q)^4] \\ &\quad = 9 + \tfrac{1}{30}\,[1385R_2 + 61R_4 - 5R_6 - R_8] - \tfrac{2}{15}\,[R_2R_8 - R_4R_6], \end{aligned} \tag{5.245}$$

where R_{2i} and $\vartheta_3(0,-q)$ are determined by (5.56) and (1.1), respectively.

The 1×1 matrices (U_3), (G_5), (R_2), and (R_4), in (5.194), (5.209), (5.225), and (5.241), respectively, were also verified directly by first computing the $n = 1$ cases of the right-hand-sides of (5.184), (5.199), (5.214), and (5.230), and then comparing termwise with the $n = 1$ cases of the right-hand-sides of (5.192), (5.207), (5.223), and (5.239), respectively.

The rest of our χ_n determinant identities involve the classical theta function $\vartheta_2(0,q)$ defined by (1.2). The entries in the $\chi_n(\{g_\nu\})$ determinant are Lambert series. Motivated by Theorems 5.11, 5.13, 5.14, 5.16, and 5.17 we exhibit these entries explicitly. We utilize the notation $\det(M_n)$ and list the Lambert series that appear in the i-th row of $\chi_n(\{g_\nu\})$.

We first have the following theorem.

Theorem 5.31. *Let $\vartheta_2(0,q)$ be defined by* (1.2), *and let $n = 1, 2, 3, \ldots$. We then have*

$$\vartheta_2(0,q)^{4n^2}\left[1+24\sum_{r=1}^{\infty}\frac{rq^{2r}}{1+q^{2r}}\right]$$

$$=\left\{4^{n(n+1)}\frac{3}{n(4n^2-1)}\prod_{r=1}^{2n-1}(r!)^{-1}\right\}\cdot\det(M_n), \qquad (5.246)$$

where M_n is the $n \times n$ matrix whose i-th row is

$$C_{2i-1}, C_{2(i+1)-1}, \ldots, C_{2(i+n-2)-1}, C_{2(i+n)-1}, \quad \textit{for } i = 1, 2, \ldots, n, \qquad (5.247)$$

when $n \geq 2$. If $n = 1$, then M_n is the 1×1 matrix (C_3). The C_{2i-1} are defined by (5.95).

Proof: Our analysis deals with $\mathrm{sd}(u,k)\,\mathrm{cn}(u,k)$. Starting with (2.83), applying row operations and (3.68), appealing to (4.44), and then utilizing (5.11) and (5.14) we have the following computation.

$$\chi_n(\{C_{2\nu-1}(q)\}) \qquad (5.248)$$

$$=\chi_n\left(\left\{(-1)^{\nu-1}\frac{z^{2\nu}k^2}{2^{2\nu+2}}\cdot(sc/d)_\nu(k^2)\right\}\right) \qquad (5.249)$$

$$=\left\{4^{-n(n+1)}\frac{n(4n^2-1)}{3}\prod_{r=1}^{2n-1}r!\right\}\cdot\tfrac{1}{2}k^{2n^2}(2-k^2)z^{2n^2+2} \qquad (5.250)$$

$$=\left\{4^{-n(n+1)}\frac{n(4n^2-1)}{3}\prod_{r=1}^{2n-1}r!\right\} \qquad (5.251a)$$

$$\cdot\,\vartheta_2(0,q)^{4n^2}\left[1+24\sum_{r=1}^{\infty}\frac{rq^{2r}}{1+q^{2r}}\right]. \qquad (5.251b)$$

Equating (5.248) with (5.251) and then solving for (5.251b) yields (5.246). □

The $n = 1$ case of Theorem 5.31 is equivalent to the identity in [259, Eq. (4) of Table 1 (vii), p. 199].

We next have the theorem.

Theorem 5.32. *Let $\vartheta_2(0,q)$ be defined by (1.2), and let $n = 1, 2, 3, \ldots$. We then have*

$$\vartheta_2(0, q^{1/2})^{4n(n+1)} \left[1 + 24 \sum_{r=1}^{\infty} \frac{rq^r}{1+q^r} \right]$$

$$= \left\{ 2^{n(4n+5)} \frac{6}{n(n+1)(2n+1)} \prod_{r=1}^{2n} (r!)^{-1} \right\} \cdot \det(M_n), \tag{5.252}$$

where M_n is the $n \times n$ matrix whose i-th row is

$$D_{2i+1}, D_{2(i+1)+1}, \ldots, D_{2(i+n-2)+1}, D_{2(i+n)+1}, \quad \textit{for } i = 1, 2, \ldots, n, \tag{5.253}$$

when $n \geq 2$. If $n = 1$, then M_n is the 1×1 matrix (D_5). The D_{2i+1} are defined by (5.96).

Proof: Our analysis deals with $\mathrm{sn}^2(u,k)$. Starting with (2.84), applying row operations and (3.68), appealing to (4.42), and then utilizing (5.12) and (5.13) we have the following computation.

$$\chi_n(\{D_{2\nu+1}(q)\}) \tag{5.254}$$

$$= \chi_n\left(\left\{(-1)^{\nu-1}\frac{z^{2\nu+2}k^2}{2^{2\nu+3}} \cdot (sn^2)_\nu(k^2)\right\}\right) \tag{5.255}$$

$$= \left\{ 2^{-n(4n+5)} \frac{n(n+1)(2n+1)}{6} \prod_{r=1}^{2n} r! \right\} \cdot 4^{n(n+1)} k^{n(n+1)} (1+k^2) z^{2n^2+2n+2} \tag{5.256}$$

$$= \left\{ 2^{-n(4n+5)} \frac{n(n+1)(2n+1)}{6} \prod_{r=1}^{2n} r! \right\} \tag{5.257a}$$

$$\cdot\, \vartheta_2(0, q^{1/2})^{4n(n+1)} \left[1 + 24 \sum_{r=1}^{\infty} \frac{rq^r}{1+q^r} \right]. \tag{5.257b}$$

Equating (5.254) with (5.257) and then solving for (5.257b) yields (5.252). □

The $n = 1$ case of Theorem 5.32 is equivalent to the identity in [259, Eq. (5) of Table 1(iii), p. 198].

Just as in (5.106), it is sometimes useful to rewrite (5.252) by utilizing (5.105). In addition, the relation (5.18) applied to the left-hand sides of (5.246) and (5.252) yields identities analogous to those in Corollary 5.12.

We next consider the χ_n determinant identities related to $\mathrm{cn}(u,k)$, $\mathrm{sn}(u,k)\,\mathrm{dn}(u,k)$, $\mathrm{dn}(u,k)$, and $\mathrm{sn}(u,k)\,\mathrm{cn}(u,k)$.

We first have the following theorem.

Theorem 5.33. *Let $\vartheta_2(0,q)$ and $\vartheta_3(0,q)$ be defined by* (1.2) *and* (1.1), *respectively. Let* $n = 1, 2, 3, \ldots$. *We then have*

$$\vartheta_2(0,q)^{2n^2}\vartheta_3(0,q)^{2n(n-1)} \cdot \left\{2n\left[1+24\sum_{r=1}^{\infty}\frac{rq^r}{1+q^r}\right]+\left[1-24\sum_{r=1}^{\infty}\frac{(2r-1)q^{2r-1}}{1+q^{2r-1}}\right]\right\}$$
$$= \left\{4^{n\frac{3}{n(2n-1)}}\prod_{r=1}^{n-1}(2r)!^{-2}\right\}\cdot \det(M_n), \tag{5.258}$$

where M_n is the $n\times n$ matrix whose i-th row is

$$T_{2i-2}, T_{2(i+1)-2}, \ldots, T_{2(i+n-2)-2}, T_{2(i+n)-2}, \quad \text{for } i = 1,2,\ldots,n, \tag{5.259}$$

when $n \geq 2$. If $n = 1$, then M_n is the 1×1 matrix (T_2). The T_{2i-2} are defined by (5.110).

Proof: Our analysis deals with $\mathrm{cn}(u,k)$. Starting with (2.85), applying row operations and (3.69), appealing to (4.33), and then utilizing (5.9), (5.11), (5.13), and (5.15) we have the following computation.

$$\chi_n(\{T_{2v-2}(q)\}) \tag{5.260}$$
$$= \chi_n\left(\left\{(-1)^{v-1}\frac{z^{2v-1}k}{4}\cdot(cn)_{v-1}(k^2)\right\}\right) \tag{5.261}$$
$$= \left\{4^{-n\frac{n(2n-1)}{3}}\prod_{r=1}^{n-1}(2r)!^2\right\}\cdot k^{n^2}[2n(1+k^2)+(1-2k^2)]z^{2n^2-n+2} \tag{5.262}$$
$$= \left\{4^{-n\frac{n(2n-1)}{3}}\prod_{r=1}^{n-1}(2r)!^2\right\} \tag{5.263a}$$
$$\cdot\,\vartheta_2(0,q)^{2n^2}\vartheta_3(0,q)^{2n(n-1)} \tag{5.263b}$$
$$\cdot\left\{2n\left[1+24\sum_{r=1}^{\infty}\frac{rq^r}{1+q^r}\right]+\left[1-24\sum_{r=1}^{\infty}\frac{(2r-1)q^{2r-1}}{1+q^{2r-1}}\right]\right\}. \tag{5.263c}$$

Equating (5.260) with (5.263) and then solving for (5.263b)–(5.263c) yields (5.258). □

The $n = 1$ case of Theorem 5.33 is equivalent to the identity in [259, Eq. (4) of Table 1(xv), p. 203].

Applying the relation (5.105) to the left-hand-side of (5.258) immediately implies that

$$\vartheta_2(0,q)^{2n}\vartheta_2(0,q^{1/2})^{4n(n-1)} \cdot \left\{2n\left[1+24\sum_{r=1}^{\infty}\frac{rq^r}{1+q^r}\right]+\left[1-24\sum_{r=1}^{\infty}\frac{(2r-1)q^{2r-1}}{1+q^{2r-1}}\right]\right\}$$
$$= \left\{4^{n^2\frac{3}{n(2n-1)}}\prod_{r=1}^{n-1}(2r)!^{-2}\right\}\cdot \det(M_n), \tag{5.264}$$

where M_n is given by (5.259).

Noting that

$$[2n(1+k^2)+(1-2k^2)] = [(4n-1)-2(n-1)(1-k^2)], \tag{5.265}$$

and recalling (5.9) and (5.10), it is sometimes useful to rewrite (5.263c) as

$$[(4n-1)\vartheta_3(0,q)^4 - 2(n-1)\vartheta_3(0,-q)^4]. \tag{5.266}$$

We next have the following theorem.

Theorem 5.34. *Let $\vartheta_2(0,q)$ and $\vartheta_3(0,q)$ be defined by* (1.2) *and* (1.1), *respectively. Let $n = 1, 2, 3, \ldots$. We then have*

$$\vartheta_2(0,q)^{2n^2}\vartheta_3(0,q)^{2n(n+1)} \cdot \left\{2n\left[1+24\sum_{r=1}^{\infty}\frac{rq^r}{1+q^r}\right] - \left[1-24\sum_{r=1}^{\infty}\frac{(2r-1)q^{2r-1}}{1+q^{2r-1}}\right]\right\}$$
$$= \left\{4^n\frac{3}{n(2n+1)}\prod_{r=1}^{n}(2r-1)!^{-2}\right\}\cdot\det(M_n), \tag{5.267}$$

where M_n is the $n \times n$ matrix whose i-th row is

$$T_{2i}, T_{2(i+1)}, \ldots, T_{2(i+n-2)}, T_{2(i+n)}, \quad \textit{for } i = 1, 2, \ldots, n, \tag{5.268}$$

when $n \geq 2$. If $n = 1$, then M_n is the 1×1 matrix (T_4). The T_{2i} are determined by (5.110).

Proof: Our analysis deals with $\mathrm{sn}(u,k)\,\mathrm{dn}(u,k)$. Starting with (2.88), applying row operations and (3.68), appealing to (4.36), and then utilizing (5.9), (5.11), (5.13), and (5.15) we have the following computation.

$$\chi_n(\{T_{2\nu}(q)\}) \tag{5.269}$$
$$= \chi_n\left(\left\{(-1)^{\nu+1}\frac{z^{2\nu+1}k}{4}\cdot(sd)_\nu(k^2)\right\}\right) \tag{5.270}$$
$$= \left\{4^{-n\frac{n(2n+1)}{3}}\prod_{r=1}^{n}(2r-1)!^2\right\}\cdot k^{n^2}[2n(1+k^2)-(1-2k^2)]z^{2n^2+n+2} \tag{5.271}$$
$$= \left\{4^{-n\frac{n(2n+1)}{3}}\prod_{r=1}^{n}(2r-1)!^2\right\} \tag{5.272a}$$
$$\cdot\,\vartheta_2(0,q)^{2n^2}\vartheta_3(0,q)^{2n(n+1)} \tag{5.272b}$$
$$\cdot\left\{2n\left[1+24\sum_{r=1}^{\infty}\frac{rq^r}{1+q^r}\right] - \left[1-24\sum_{r=1}^{\infty}\frac{(2r-1)q^{2r-1}}{1+q^{2r-1}}\right]\right\}. \tag{5.272c}$$

Equating (5.269) with (5.272) and then solving for (5.272b)–(5.272c) yields (5.267).

□

The $n = 1$ case of Theorem 5.34 is equivalent to the identity in [259, Eq. (5) of Table 1(xv), p. 203].

Applying the relation (5.105) to the left-hand-side of (5.267) immediately implies that

$$\vartheta_2(0,q^{1/2})^{4n^2}\vartheta_3(0,q)^{2n}\cdot\left\{2n\left[1+24\sum_{r=1}^{\infty}\frac{rq^r}{1+q^r}\right]-\left[1-24\sum_{r=1}^{\infty}\frac{(2r-1)q^{2r-1}}{1+q^{2r-1}}\right]\right\}$$
$$=\left\{4^{n(n+1)}\frac{3}{n(2n+1)}\prod_{r=1}^{n}(2r-1)!^{-2}\right\}\cdot\det(M_n), \tag{5.273}$$

where M_n is given by (5.268).

Equation (5.273) can be rewritten as in Corollary 5.15 by utilizing (5.18), with $q \mapsto q^{1/2}$. Noting that

$$[2n(1+k^2)-(1-2k^2)] = [(4n+1)-2(n+1)(1-k^2)], \tag{5.274}$$

and recalling (5.9) and (5.10), it is sometimes useful to rewrite (5.272c) as

$$[(4n+1)\vartheta_3(0,q)^4-2(n+1)\vartheta_3(0,-q)^4]. \tag{5.275}$$

We next have the following theorem.

Theorem 5.35. *Let $\vartheta_2(0,q)$ and $\vartheta_3(0,q)$ be defined by* (1.2) *and* (1.1), *respectively. Let $n = 1, 2, 3, \ldots$. We then have*

$$\vartheta_2(0,q)^{2n(n-1)}\vartheta_3(0,q)^{2n^2}\cdot\left\{n\left[1+24\sum_{r=1}^{\infty}\frac{rq^r}{1+q^r}\right]-\left[1+24\sum_{r=1}^{\infty}\frac{rq^{2r}}{1+q^{2r}}\right]\right\}$$
$$=\left\{4^{n^2}\frac{3}{2n(2n-1)}\prod_{r=1}^{n-1}(2r)!^{-2}\right\}\cdot\det(M_n) \tag{5.276a}$$
$$+\left\{4^{n^2-1}\frac{3}{2n(2n-1)}\prod_{r=1}^{n-1}(2r)!^{-2}\right\}\cdot\det(\bar{M}_{n-1}), \tag{5.276b}$$

where the matrices M_n and $\bar{M}_{n-1}$ are defined as follows:

The $n \times n$ matrix M_n has i-th row given by

$$N_{2i-2}, N_{2(i+1)-2}, \ldots, N_{2(i+n-2)-2}, N_{2(i+n)-2}, \tag{5.277}$$

for $i = 1, 2, \ldots, n$, when $n \geq 2$. If $n = 1$, then M_n is the 1×1 matrix (N_2). The N_{2i-2} are defined by (5.128).

The $(n-1) \times (n-1)$ matrix $\bar{M}_{n-1}$ has i-th row given by

$$N_{2i+2}, N_{2(i+1)+2}, \ldots, N_{2(i+n-3)+2}, N_{2(i+n-1)+2}, \tag{5.278}$$

for $i = 1, 2, \ldots, n-1$, when $n \geq 3$. If $n = 1$, then $\bar{M}_{n-1}$ is defined to be 0. If $n = 2$, then $\bar{M}_{n-1}$ is the 1×1 matrix (N_6). The N_{2i-2} are defined by (5.128).

Proof: Our analysis deals with $\mathrm{dn}(u,k)$. Starting with (2.86) and (2.87), solving for $(-1)^m(z^{2m+1}/2^{2m+2})\cdot(dn)_m(k^2)$ for $m\geq 0$, applying row operations, recalling (3.56), and simplifying, leads to the following identity.

$$\chi_n(\{N_{2\nu-2}(q)\})+\tfrac{1}{4}\chi_{n-1}(\{N_{2\nu+2}(q)\}) \tag{5.279}$$

$$=\chi_n\left(\left\{(-1)^{\nu-1}\frac{z^{2\nu-1}}{4^\nu}\cdot(dn)_{\nu-1}(k^2)\right\}\right). \tag{5.280}$$

Next, applying row operations and (3.69) to (5.280), appealing to (4.34), and then utilizing (5.9), (5.11), (5.13), and (5.14) we obtain

$$=\left\{4^{-n^2}\frac{n(2n-1)}{3}\prod_{r=1}^{n-1}(2r)!^2\right\}\cdot k^{n(n-1)}\left[2n(1+k^2)-(2-k^2)\right]z^{2n^2-n+2} \tag{5.281}$$

$$=\left\{4^{-n^2}\frac{n(2n-1)}{3}\prod_{r=1}^{n-1}(2r)!^2\right\} \tag{5.282a}$$

$$\cdot\vartheta_2(0,q)^{2n(n-1)}\vartheta_3(0,q)^{2n^2} \tag{5.282b}$$

$$\cdot\left\{2n\left[1+24\sum_{r=1}^{\infty}\frac{rq^r}{1+q^r}\right]-2\left[1+24\sum_{r=1}^{\infty}\frac{rq^{2r}}{1+q^{2r}}\right]\right\}. \tag{5.282c}$$

Equating (5.279) with (5.282) and then solving for (5.282b)–(5.282c) yields (5.276). □

The $n=1$ case of Theorem 5.35 is equivalent to the identity in [259, Eq. (4) of Table 1(xi), p. 201]. Here, the 0×0 determinant in (5.276b) is defined to be 0.

Noting that

$$[2n(1+k^2)-(2-k^2)]=[(4n-1)-(2n+1)(1-k^2)], \tag{5.283}$$

and recalling (5.9) and (5.10), it is sometimes useful to rewrite (5.282c) as

$$[(4n-1)\vartheta_3(0,q)^4-(2n+1)\vartheta_3(0,-q)^4]. \tag{5.284}$$

We next have the following theorem.

Theorem 5.36. *Let $\vartheta_2(0,q)$ and $\vartheta_3(0,q)$ be defined by* (1.2) *and* (1.1), *respectively. Let $n=1,2,3,\ldots$. We then have*

$$\vartheta_2(0,q)^{2n(n+1)}\vartheta_3(0,q)^{2n^2}\cdot\left\{n\left[1+24\sum_{r=1}^{\infty}\frac{rq^r}{1+q^r}\right]+\left[1+24\sum_{r=1}^{\infty}\frac{rq^{2r}}{1+q^{2r}}\right]\right\}$$

$$=\left\{4^{n(n+1)}\frac{6}{n(2n+1)}\prod_{r=1}^{n}(2r-1)!^{-2}\right\}\cdot\det(M_n), \tag{5.285}$$

where M_n is the $n \times n$ matrix whose i-th row is

$$N_{2i}, N_{2(i+1)}, \ldots, N_{2(i+n-2)}, N_{2(i+n)}, \quad \text{for } i = 1, 2, \ldots, n, \tag{5.286}$$

when $n \geq 2$. If $n = 1$, then M_n is the 1×1 matrix (N_4). The N_{2i} are determined by (5.128).

Proof: Our analysis deals with $\mathrm{sn}(u,k)\ \mathrm{cn}(u,k)$. Starting with (2.89), applying row operations and (3.68), appealing to (4.35), and then utilizing (5.9), (5.11), (5.13), and (5.14) we have the following computation.

$$\chi_n(\{N_{2\nu}(q)\}) \tag{5.287}$$

$$= \chi_n\left(\left\{(-1)^{\nu+1}\frac{z^{2\nu+1}k^2}{2^{2\nu+2}} \cdot (sc)_\nu(k^2)\right\}\right) \tag{5.288}$$

$$= \left\{4^{-(n^2+n+1)}\frac{n(2n+1)}{3}\prod_{r=1}^{n}(2r-1)!^2\right\} \cdot k^{n(n+1)}[2n(1+k^2)+(2-k^2)]z^{2n^2+n+2} \tag{5.289}$$

$$= \left\{4^{-(n^2+n+1)}\frac{n(2n+1)}{3}\prod_{r=1}^{n}(2r-1)!^2\right\} \tag{5.290a}$$

$$\cdot\, \vartheta_2(0,q)^{2n(n+1)}\vartheta_3(0,q)^{2n^2} \tag{5.290b}$$

$$\cdot \left\{2n\left[1+24\sum_{r=1}^{\infty}\frac{rq^r}{1+q^r}\right]+2\left[1+24\sum_{r=1}^{\infty}\frac{rq^{2r}}{1+q^{2r}}\right]\right\}. \tag{5.290c}$$

Equating (5.287) with (5.290) and then solving for (5.290b)–(5.290c) yields (5.285).

□

The $n = 1$ case of Theorem 5.36 is equivalent to the identity in [259, Eq. (5) of Table 1(xi), p. 201].

Applying the relation (5.105) to the left-hand-side of (5.285) immediately implies that

$$\vartheta_2(0,q^{1/2})^{4n^2}\vartheta_2(0,q)^{2n} \cdot \left\{n\left[1+24\sum_{r=1}^{\infty}\frac{rq^r}{1+q^r}\right]+\left[1+24\sum_{r=1}^{\infty}\frac{rq^{2r}}{1+q^{2r}}\right]\right\}$$

$$= \left\{4^{n(2n+1)}\frac{6}{n(2n+1)}\prod_{r=1}^{n}(2r-1)!^{-2}\right\} \cdot \det(M_n), \tag{5.291}$$

where M_n is given by (5.286).

Equation (5.291) can be rewritten as in Corollary 5.15 by utilizing (5.18), with $q \mapsto q^{1/2}$. Noting that

$$[2n(1+k^2)+(2-k^2)] = [(4n+1)-(2n-1)(1-k^2)], \tag{5.292}$$

and recalling (5.9) and (5.10), it is sometimes useful to rewrite (5.290c) as

$$[(4n+1)\vartheta_3(0,q)^4-(2n-1)\vartheta_3(0,-q)^4]. \tag{5.293}$$

By appealing to the infinite product expansions for the classical theta functions in [250, pp. 472–473] and [6, Eq. (10.7.8), p. 510] our expansion formulas in this section for powers of various products of classical theta functions is transformed into expansions for the corresponding infinite products.

We close this section with some more detailed observations about the formulas for $r_{16}(n)$ and $r_{24}(n)$ in (1.26) and (1.28), respectively. We also provide more information about the analysis of the formulas for $r_{4n^2}(N)$ and $r_{4n(n+1)}(N)$ obtained by taking the coefficient of q^N in Theorems 5.4 and 5.6. This analysis is based upon the elementary estimates for divisor sums in [94, pp. 122–123; p. 125] and the estimate given by the following lemma.

Lemma 5.37. *Let $n = 1, 2, 3, \ldots$ and let $b_1, b_2, \ldots, b_n$ be fixed nonnegative integers. Let $N \geq n$ be an integer. Then, there exists positive constants c_1 and c_2, independent of N, such that*

$$c_1 N^{b_1+\cdots+b_n+n-1} \leq \sum_{\substack{m_1+\cdots+m_n=N \\ m_i \geq 1}} m_1^{b_1} m_2^{b_2} \ldots m_n^{b_n} \leq c_2 N^{b_1+\cdots+b_n+n-1}. \tag{5.294}$$

Proof: Lemma 5.37 follows by induction from the $n = 2$ case, which in turn is a consequence of the Euler-Maclaurin sum formula [189, Eq. (7.2), p. 14], a simple case of the beta integral, and several applications of the binomial theorem. □

We first study (1.26) and (1.28). Equation (1.28b) is the dominate term for $r_{24}(n)$, and (1.28a) is the "remainder term" of lower order of magnitude. The elementary analysis in [94, pp. 122–123] immediately implies that

$$(2 - \zeta(r))n^r \leq (-1)^n \sigma_r^\dagger(n) \leq \zeta(r) n^r, \tag{5.295}$$

where $\sigma_r^\dagger(n)$ is given by (1.14), $\zeta(r)$ is the Riemann ζ-function, and $r \geq 2$. Applying (5.295), and then Lemma 5.37 as needed, to (1.28) it follows from computer algebra [251] computations that

$$0.0120n^{11} < \alpha_{24}(n) < 0.0332n^{11}, \tag{5.296}$$

and

$$3.52587n^7 < \beta_{24}(n) < 36.3288n^3 + 14.7474n^5 + 3.58524n^7, \tag{5.297}$$

for $n \geq 3$, where

$$r_{24}(n) = \alpha_{24}(n) + \beta_{24}(n), \tag{5.298}$$

with $\alpha_{24}(n)$ and $\beta_{24}(n)$ given by (1.28b) and (1.28a), respectively. The upper bound in (5.297) may be replaced by $\beta_{24}(n) < 3.59n^7$, when $n \geq 56$. Adding the bounds in (5.296) and (5.297) gives upper and lower bounds for $r_{24}(n)$ that are consistent with [94, Eq. (9.20), p. 122]. More computer algebra suggests that $\alpha_{24}(n)$, and hence $r_{24}(n)$, is very close to $0.0231n^{11}$. Moreover, for n large, it appears that $\beta_{24}(n)$ is very close to either $3.527n^7$,

$3.529n^7$, $3.555n^7$, $3.557n^7$, $3.583n^7$, $3.584n^7$, or $3.585n^7$, depending on the congruence class of n. Noting that $\beta_{24}(1) = 48$ and $\beta_{24}(2) = 1104$ we have $\beta_{24}(n) > 0$, for $n \geq 1$. Similarly, $\alpha_{24}(1) = \alpha_{24}(2) = 0$, and $\alpha_{24}(n) > 0$, for $n \geq 3$. The power series form of (1.22) given by Corollary 8.2 implies that $\alpha_{24}(n)$ and $\beta_{24}(n)$ are integers for $n \geq 1$. Just do a termwise analysis of (8.2b) which considers the congruence classes mod 3 of m_1 and m_2, and a termwise analysis of (8.2a) with the congruence classes mod 9 of m_1.

Equation (1.26b) is the dominate term for $r_{16}(n)$, and (1.26a) is the "remainder term" of lower order of magnitude, at least for all sufficiently large n. We suspect this is true for all $n \geq 1$. The elementary analysis in [94, p. 125] involving $r_{12}(n)$ immediately implies that

$$(2 - \zeta(r))n^r \leq (-1)^{n+1}\sigma_r^{\sim}(n) \leq \zeta(r)n^r, \tag{5.299}$$

where $\sigma_r^{\sim}(n)$ is given by (1.27), $\zeta(r)$ is the Riemann ζ-function, and $r \geq 2$. We also have

$$(2 - \mathcal{H}_n)n \leq (-1)^{n+1}\sigma_1^{\sim}(n) \leq n\mathcal{H}_n, \tag{5.300}$$

where $\mathcal{H}_n$ is the harmonic number given by

$$\mathcal{H}_n := \sum_{p=1}^{n} \frac{1}{p}. \tag{5.301}$$

Maximizing the standard estimate for $\mathcal{H}_n$ in [189, Eq. (16.1), p. 28] with $m = 2$, and then applying (5.299) and (5.300) to (1.26a), we obtain by computer algebra that

$$10.2728n^5 < \beta_{16}(n) < 11.0650n^5, \tag{5.302}$$

for $n \geq 55$, where

$$r_{16}(n) = \alpha_{16}(n) + \beta_{16}(n), \tag{5.303}$$

with $\alpha_{16}(n)$ and $\beta_{16}(n)$ given by (1.26b) and (1.26a), respectively. The lower bound in (5.302) also holds for $n \geq 1$. From the $k = 16$ case of [94, Eq. (9.20), p. 122] there are positive constants c_1 and c_2 such that

$$c_1 n^7 \leq r_{16}(n) \leq c_2 n^7, \tag{5.304}$$

for $n \geq 1$. By combining (5.302), (5.303), and (5.304) there are positive constants c_3 and c_4 such that

$$c_3 n^7 < \alpha_{16}(n) < c_4 n^7, \tag{5.305}$$

for all sufficiently large n, with $\alpha_{16}(n)$ given by (1.26b). More computer algebra suggests that $\alpha_{16}(n)$, and hence $r_{16}(n)$, is close to $1.87n^7$. Moreover, for n large, it appears that $\beta_{16}(n)$ is very close to either $10.32n^5$, $10.34n^5$, $10.36n^5$, $10.38n^5$, $10.66n^5$, $10.67n^5$, $10.71n^5$, $11.00n^5$, or $11.04n^5$, depending on the congruence class of n. The power series form of

(1.20) given by Corollary 8.1 implies that $\alpha_{16}(n)$ and $\beta_{16}(n)$ are integers for $n \geq 1$. Just do a termwise analysis of (8.1a) and (8.1b) which considers the congruence classes mod 3 of m_1 and m_2.

The analysis of the terms in the formulas for $r_{4n^2}(N)$ and $r_{4n(n+1)}(N)$ described just after the proof of Theorem 5.6 follows from the elementary estimates for divisor sums in (5.295), (5.299), and (5.300), and suitable applications of Lemma 5.37.

6. Schur functions and Lambert series

In this section we first establish a Schur function expansion of a general $n \times n$ determinant whose entries are either constants or Lambert series. This expansion applied to Section 5 yields the Schur function form of our Hankel determinant identities in Section 7 and also completes our proof of the Kac–Wakimoto conjectured identities for triangular numbers in [120, p. 452]. At the end of this section we state an analogous expansion (whose proof is the same) which leads to the Schur function form of our χ_n determinant identities in Section 7.

Throughout this section we use the notation $I_n := \{1, 2, \ldots, n\}$, $\|S\|$ is the cardinality of the set $S \subseteq I_n$, and $\det(M)$ is the determinant of the $n \times n$ matrix M. We frequently consider the sets S and T, with

$$S := \{\ell_1 < \ell_2 < \cdots < \ell_p\} \quad \text{and} \quad S^c := \{\ell_{p+1} < \cdots < \ell_n\}, \tag{6.1}$$

$$T := \{j_1 < j_2 < \cdots < j_p\} \quad \text{and} \quad T^c := \{j_{p+1} < \cdots < j_n\}, \tag{6.2}$$

where $S^c := I_n - S$ is the compliment of the set S, and $p = 1, 2, \ldots, n$. We also have

$$\Sigma(S) := \ell_1 + \ell_2 + \cdots + \ell_p \quad \text{and} \quad \Sigma(T) := j_1 + j_2 + \cdots + j_p. \tag{6.3}$$

The Lambert series entries in our determinants are of the form given by the following definition.

Definition 6.1 (A general Lambert series). Let A, B, C, D, E, F, G, and u be indeterminant, and assume that $0 < |q| < 1$. Then, we define the Lambert series L_u by

$$\begin{aligned} L_u &\equiv L_u(m_1; A, \ldots, G) \equiv L_u(m_1; A, B, C, D, E, F, G) \\ &:= \sum_{m_1=1}^{\infty} \frac{D^u E^{m_1} (Bm_1 + C)^u}{1 + Aq^{Bm_1+C}} \cdot q^{Fm_1+G} && (6.4) \\ &= (-A)^{-1} q^{G-C} \sum_{y_1, m_1 \geq 1} (-A)^{y_1} E^{m_1} (D(Bm_1 + C))^u \cdot q^{(F-B)m_1} q^{(Bm_1+C)y_1}. && (6.5) \end{aligned}$$

We transform (6.4) into (6.5) by expanding $1/(1 + Aq^{Bm_1+C})$ as a geometric series and then interchanging summation.

Our aim is to obtain a Schur function expansion of the $n \times n$ determinant $\det(M_{n,S,\{b_r\},\{c_r\},\{a_r\}})$ of the matrix given by the following definition.

Definition 6.2 (First Lambert series matrix). Let $b_1, b_2, \ldots, b_n$ and $c_1, c_2, \ldots, c_n$ be indeterminant, and let $n = 1, 2, \ldots$. Furthermore, assume that $b_1 < b_2 < \cdots < b_n$ and $c_1 < c_2 < \cdots < c_n$. Take $\{a_r : r = 1, 2, \ldots\}$ to be an arbitrary sequence. Let S and S^c be the subsets of I_n in (6.1), with $p = 1, 2, \ldots, n$. Let $L_u(r; A, \ldots, G)$ be the Lambert series in Definition 6.1. Then,

$$M_{n,S,\{b_r\},\{c_r\},\{a_r\}} \equiv M_{n,S,\{b_1,\ldots,b_n\},\{c_1,\ldots,c_n\},\{a_r\}} \tag{6.6}$$

is defined to be the $n \times n$ matrix whose i-th row is

$$L_{c_i+b_1}(m_\mu; A, \ldots, G), L_{c_i+b_2}(m_\mu; A, \ldots, G), \ldots, L_{c_i+b_n}(m_\mu; A, \ldots, G),$$

if $i = \ell_\mu \in S$, and

$$a_i, a_{i+1}, \ldots, a_{i+n-1}, \quad \text{if } i \notin S. \tag{6.7}$$

In order to state our expansion formula we first need a few more definitions.

Definition 6.3 (Divisors). Let $b_1 < b_2 < \cdots < b_n$ and $c_1 < c_2 < \cdots < c_n$ be integers. Then define the divisors d_b and d_c by

$$d_b := \text{any common divisor of } \{b_r - b_1 \mid 2 \leq r \leq n\}, \tag{6.8}$$

$$d_c := \text{any common divisor of } \{c_s - c_r \mid 1 \leq r < s \leq n\}. \tag{6.9}$$

We often use the greatest common divisor.

Definition 6.4 (First Laplace expansion formula determinant). Let $\{a_r : r = 1, 2, \ldots\}$ be an arbitrary sequence. Let the sets S^c and T^c be given by (6.1) and (6.2), and let $p = 1, 2, \ldots, n$. Then,

$$\det\left(D_{n-p,S^c,T^c}\right) \tag{6.10}$$

is the determinant of the $(n-p) \times (n-p)$ matrix

$$D_{n-p,S^c,T^c} := \left[a_{(\ell_{p+r}+j_{p+s}-1)}\right]_{1 \leq r,s \leq n-p}. \tag{6.11}$$

Definition 6.5 (Schur functions). Let $\lambda = (\lambda_1, \lambda_2, \ldots, \lambda_i, \ldots)$ be a partition of nonnegative integers in decreasing order, $\lambda_1 \geq \lambda_2 \geq \cdots \geq \lambda_i \ldots$, such that only finitely many of the λ_i are nonzero. The length $\ell(\lambda)$ is the number of nonzero parts of λ. Given a partition $\lambda = (\lambda_1, \lambda_2, \ldots, \lambda_p)$ of length $\leq p$,

$$s_\lambda(x) \equiv s_\lambda(x_1, x_2, \ldots, x_p) := \frac{\det\left(x_i^{\lambda_j+p-j}\right)}{\det\left(x_i^{p-j}\right)} \tag{6.12}$$

is the Schur function [153] corresponding to the partition λ. (Here, $\det(a_{ij})$ denotes the determinant of a $p \times p$ matrix with (i, j)-th entry a_{ij}). The Schur function $s_\lambda(x)$ is a symmetric polynomial in $x_1, x_2, \ldots, x_p$ with nonnegative integer coefficients. We typically have $p = 1, 2, \ldots, n$.

Definition 6.6 (Expansion formula partitions). Let $b_1, b_2, \ldots, b_n$ and $c_1, c_2, \ldots, c_n$ be indeterminant. Let $n = 1, 2, \ldots$ and $p = 1, 2, \ldots, n$. Furthermore, assume that $b_1 < b_2 < \cdots < b_n$ and $c_1 < c_2 < \cdots < c_n$. Let d_b and d_c be determined by Definition 6.3. Take $\{\ell_1, \ell_2, \ldots, \ell_p\}$ and $\{j_1, j_2, \ldots, j_p\}$ as in (6.1) and (6.2). We then define the partitions $\lambda = (\lambda_1, \lambda_2, \ldots, \lambda_p)$ and $\nu = (\nu_1, \nu_2, \ldots, \nu_p)$ of length $\leq p$ as follows:

$$\lambda_i := \tfrac{1}{d_c} c_{\ell_{p-i+1}} - \tfrac{1}{d_c} c_{\ell_1} + i - p, \quad \text{for } i = 1, 2, \ldots, p, \tag{6.13}$$

$$\nu_i := \tfrac{1}{d_b} b_{j_{p-i+1}} - \tfrac{1}{d_b} b_{j_1} + i - p, \quad \text{for } i = 1, 2, \ldots, p. \tag{6.14}$$

Keeping in mind Definitions 6.1 to 6.6 we now have the following theorem.

Theorem 6.7 (*Expansion of first Lambert series determinant*). *Let $L_u(r; A, \ldots, G)$ be the Lambert series in Definition 6.1. Let the $n \times n$ matrix $M_{n,S,\{b_r\},\{c_r\},\{a_r\}}$ of Lambert series be given by Definition 6.2, and take $n = 1, 2, \ldots$ and $p = 1, 2, \ldots, n$. Let A, B, C, D, E, F, G, and $b_1, b_2, \ldots, b_n$ and $c_1, c_2, \ldots, c_n$ be indeterminant, and assume that $0 < |q| < 1$. Let S, T, S^c, T^c, $\Sigma(S)$, and $\Sigma(T)$ be given by (6.1)–(6.3). Take $\{a_r : r = 1, 2, \ldots\}$ to be an arbitrary sequence. Let d_b and d_c be given by Definition 6.3 and s_λ and s_ν be the Schur functions in Definition 6.5 with partitions λ and ν as in Definition 6.6. Finally, take the $(n-p) \times (n-p)$ matrix D_{n-p,S^c,T^c} in Definition 6.4. We then have the expansion formula*

$$\begin{aligned}
\det\left(M_{n,S,\{b_r\},\{c_r\},\{a_r\}}\right) = (-A)^{-p} q^{p(G-C)} & \sum_{\substack{y_1,\ldots,y_p \geq 1 \\ m_1 > m_2 > \cdots > m_p \geq 1}} (-A)^{y_1+\cdots+y_p} E^{m_1+\cdots+m_p} \\
& \cdot q^{(F-B)(m_1+\cdots+m_p)} q^{(Bm_1+C)y_1+\cdots+(Bm_p+C)y_p} \\
& \cdot \prod_{1\leq r<s\leq p} \left((D(Bm_r+C))^{d_c} - (D(Bm_s+C))^{d_c}\right) \\
& \cdot \prod_{1\leq r<s\leq p} \left((D(Bm_r+C))^{d_b} - (D(Bm_s+C))^{d_b}\right) \\
& \cdot s_\lambda\left((D(Bm_1+C))^{d_c}, \ldots, (D(Bm_p+C))^{d_c}\right) \\
& \cdot \sum_{\substack{\emptyset \subset T \subseteq I_n \\ \|T\|=p}} (-1)^{\Sigma(S)+\Sigma(T)} \cdot \det\left(D_{n-p,S^c,T^c}\right) \cdot \left(\prod_{r=1}^{p} D(Bm_r+C)\right)^{c_{\ell_1}+b_{j_1}} \\
& \cdot s_\nu\left((D(Bm_1+C))^{d_b}, \ldots, (D(Bm_p+C))^{d_b}\right).
\end{aligned} \tag{6.15}$$

We give a derivation proof of Theorem 6.7 that relies upon the Laplace expansion formula for a determinant, classical properties of Schur functions, symmetry and skew-symmetry arguments, and row and column operations.

We start with the Laplace expansion formula for a determinant as given in [118, pp. 396–397].

Theorem 6.8 (*Laplace expansion formula for a determinant*). *Let Λ be an $n \times n$ matrix with $n = 1, 2, \ldots$. Let S, T, S^c, T^c, $\Sigma(S)$, and $\Sigma(T)$ be given by* (6.1)–(6.3), *where $I_n := \{1, 2, \ldots, n\}$, $S^c := I_n - S$ is the compliment of the set S, and $p = 1, 2, \ldots, n$. Let $\Lambda_{S,T}$ denote the minor of Λ obtained from the rows and columns of S and T, respectively. Similarly, let Λ_{S^c,T^c} be the minor corresponding to S^c and T^c. We then expand the determinant of Λ along a given fixed set $S = \{\ell_1 < \ell_2 < \cdots < \ell_p\}$ of rows as follows:*

$$\det(\Lambda) = \sum_{\substack{T \subseteq I_n \\ \|T\|=p}} \Lambda_{S,T} \Lambda_{S^c,T^c} \cdot (-1)^{\Sigma(S)+\Sigma(T)}. \tag{6.16}$$

If we successively expand $\det(M_{n,S,\{b_r\},\{c_r\},\{a_r\}})$ along those rows of the a_r's that correspond to $i \in S^c$, we obtain a linear combination of $p \times p$ determinants of the L_b's. To this end, apply Theorem 6.8, where $\Lambda_{S,T}$ is a $p \times p$ minor involving Lambert series, and Λ_{S^c,T^c} depends upon the constants $\{a_r : r = 1, 2, \ldots\}$. We have the following expansion.

$$\det\left(M_{n,S,\{b_r\},\{c_r\},\{a_r\}}\right) = \sum_{\substack{T \subseteq I_n \\ \|T\|=p}} \beta_{S,T} \cdot \det\left(L_{(S,T,r,s)}\right)_{1 \le r,s \le p}, \tag{6.17}$$

where $S \subseteq I_n$ is fixed, $\beta_{S,T}$ are constants, (S, T, r, s) is an integer that depends on S, T, r, and s, and $L_{(S,T,r,s)}$ is a Lambert series from Definition 6.1.

In the analysis that follows, we need an expansion closely related to (6.17). If we replace $L_{(S,T,r,s)}$ in the right-hand-side of (6.17) by the simpler expression $(D(Bm_r + C))^{(S,T,r,s)}$, and then reverse the steps (via Theorem 6.8) that led to (6.17), we immediately obtain the following lemma.

Lemma 6.9 (*Umbral calculus trick*). *Let the constants $\beta_{S,T}$ and integers (S, T, r, s) be determined by* (6.17). *Let S, $\{b_r\}$, $\{c_r\}$, $\{a_r\}$ satisfy the conditions in Definition* 6.2. *Take $n = 1, 2, \ldots$ and $p = 1, 2, \ldots, n$. We then have the expansion*

$$\det\left(P_{n,S,\{b_r\},\{c_r\},\{a_r\}}\right) = \sum_{\substack{T \subseteq I_n \\ \|T\|=p}} \beta_{S,T} \cdot \det\left((D(Bm_r + C))^{(S,T,r,s)}\right)_{1 \le r,s \le p}, \tag{6.18}$$

where

$$P_{n,S,\{b_r\},\{c_r\},\{a_r\}} \equiv P_{n,S,\{b_1,\ldots,b_n\},\{c_1,\ldots,c_n\},\{a_r\}} \tag{6.19}$$

is defined to be the $n \times n$ matrix whose i-th row is

$$(D(Bm_\mu + C))^{c_i+b_1}, (D(Bm_\mu + C))^{c_i+b_2}, \ldots, (D(Bm_\mu + C))^{c_i+b_n},$$

if $i = \ell_\mu \in S$, *and*

$$a_i, a_{i+1}, \ldots, a_{i+n-1}, \quad \textit{if } i \notin S. \tag{6.20}$$

It will also be useful to consider the determinant of (6.19) as a function of $\{m_1, m_2, \ldots, m_p\}$. That is,

$$f(m_1, \ldots, m_p) := \det\left(P_{n,S,\{b_r\},\{c_r\},\{a_r\}}\right). \tag{6.21}$$

Then, if $\rho \in \mathcal{S}_p$ is any permutation of $\{1, 2, \ldots, p\}$, we have from (6.18) that

$$f\left(m_{\rho(1)}, \ldots, m_{\rho(p)}\right) = \sum_{\substack{T \subseteq I_n \\ \|T\|=p}} \beta_{S,T} \cdot \det\left((D(Bm_{\rho(r)} + C))^{(S,T,r,s)}\right)_{1 \le r,s \le p}. \tag{6.22}$$

As a first step in transforming (6.17) we write the determinant on the right-hand-side as

$$\det\left(L_{(S,T,r,s)}\right)_{1 \le r,s \le p} = \sum_{\sigma \in \mathcal{S}_p} \operatorname{sign}(\sigma) \prod_{r=1}^{p} L_{(S,T,r,\sigma(r))}, \tag{6.23}$$

where we pick a different element from each row.

We expand each product on the right-hand-side of (6.23) into a double multiple sum by appealing to the following lemma.

Lemma 6.10. *Let* $b_1, b_2, \ldots, b_p$ *be indeterminant, and let* $p = 1, 2, \ldots$. *Let* $L_u(r; A, \ldots, G)$ *be the Lambert series in Definition* 6.1. *Let* $\rho \in \mathcal{S}_p$ *be any fixed permutation of* $\{1, 2, \ldots, p\}$. *Then,*

$$\begin{aligned}\prod_{r=1}^{p} L_{b_r} &\equiv \prod_{r=1}^{p} L_{b_r}(m_r; A, B, C, D, E, F, G) \\ &= (-A)^{-p} q^{p(G-C)} \sum_{\substack{y_1, \ldots, y_p \ge 1 \\ m_1, \ldots, m_p \ge 1}} (-A)^{y_1 + \cdots + y_p} E^{m_1 + \cdots + m_p} \\ &\quad \cdot q^{(F-B)(m_1 + \cdots + m_p)} q^{(Bm_1+C)y_1 + \cdots + (Bm_p+C)y_p} \\ &\quad \cdot \prod_{r=1}^{p} \left(D\left(Bm_{\rho(r)} + C\right)\right)^{b_r}.\end{aligned} \tag{6.24}$$

Proof: Apply (6.5), with $y_{\rho(r)}$ and $m_{\rho(r)}$, to the r-th factor in the left-hand-side of (6.24), interchange summation, note that $(Bm_1 + C)y_1 + \cdots + (Bm_p + C)y_p = B(m_1 y_1 + \cdots + m_p y_p) + C(y_1 + \cdots + y_p)$, and use symmetry. □

Applying Lemma 6.10 with the same $\rho \in S_p$ to each term in the right-hand-side of (6.23), and then interchanging summation, yields

$$\begin{aligned}\det\left(L_{(S,T,r,s)}\right)_{1\le r,s\le p} &= (-A)^{-p} q^{p(G-C)} \sum_{\substack{y_1,\dots,y_p\ge 1\\ m_1,\dots,m_p\ge 1}} (-A)^{y_1+\cdots+y_p} E^{m_1+\cdots+m_p} \\ &\quad \cdot q^{(F-B)(m_1+\cdots+m_p)} q^{(Bm_1+C)y_1+\cdots+(Bm_p+C)y_p} \\ &\quad \cdot \sum_{\sigma\in S_p} \operatorname{sign}(\sigma) \prod_{r=1}^{p} \left(D(Bm_{\rho(r)}+C)\right)^{(S,T,r,\sigma(r))} \\ &= (-A)^{-p} q^{p(G-C)} \sum_{\substack{y_1,\dots,y_p\ge 1\\ m_1,\dots,m_p\ge 1}} (-A)^{y_1+\cdots+y_p} E^{m_1+\cdots+m_p} \\ &\quad \cdot q^{(F-B)(m_1+\cdots+m_p)} q^{(Bm_1+C)y_1+\cdots+(Bm_p+C)y_p} \\ &\quad \cdot \left\{ \det\left(\left(D(Bm_r+C)\right)^{(S,T,r,s)}\right)_{1\le r,s\le p}\Big|_{m_r\to m_{\rho(r)}} \right\}. \end{aligned} \tag{6.25}$$

Remark. In Eq. (6.23) we used the definition of determinant in which a different column element is selected from each row. This simplified the above computation. Had we chosen a different row element from each column, we would have had to apply the $\rho\circ\sigma$ case of Lemma 6.10 to the σ term in (6.23).

It is now not difficult to see that (6.17)–(6.22), (6.25), and an interchange of summation yields the following lemma.

Lemma 6.11. *Let $M_{n,S,\{b_r\},\{c_r\},\{a_r\}}$ be determined by Definition 6.2 and $P_{n,S,\{b_r\},\{c_r\},\{a_r\}}$ by (6.20). Let $n = 1, 2, \ldots$ and $p = 1, 2, \ldots, n$. Let $\rho \in S_p$ be any fixed permutation of $\{1, 2, \ldots, p\}$. Then,*

$$\begin{aligned}\det\left(M_{n,S,\{b_r\},\{c_r\},\{a_r\}}\right) &= (-A)^{-p} q^{p(G-C)} \sum_{\substack{y_1,\dots,y_p\ge 1\\ m_1,\dots,m_p\ge 1}} (-A)^{y_1+\cdots+y_p} E^{m_1+\cdots+m_p} \\ &\quad \cdot q^{(F-B)(m_1+\cdots+m_p)} q^{(Bm_1+C)y_1+\cdots+(Bm_p+C)y_p} \\ &\quad \cdot \left\{ \det\left(P_{n,S,\{b_r\},\{c_r\},\{a_r\}}\right)\Big|_{m_r\to m_{\rho(r)}} \right\}. \end{aligned} \tag{6.26}$$

Before symmetrizing (6.26) with respect to $\rho \in S_p$, we transform the determinant $\det(P_{n,S,\{b_r\},\{c_r\},\{a_r\}})$ by some row operations.

For each $i = \ell_r \in S$, we factor $(D(Bm_r + C))^{c_{\ell_r}+b_1}$ out of the i-th row of the $n \times n$ matrix $P_{n,S,\{b_r\},\{c_r\},\{a_r\}}$. Keeping in mind d_b and d_c from Definition 6.3, and the simple relation $c_{\ell_r} = c_{\ell_1} + d_c(c_{\ell_r} - c_{\ell_1})/d_c$, we write $\det(P_{n,S,\{b_r\},\{c_r\},\{a_r\}})$ as follows:

$$\begin{aligned}\det\left(P_{n,S,\{b_r\},\{c_r\},\{a_r\}}\right) &= \prod_{r=1}^{p} (D(Bm_r+C))^{c_{\ell_1}+b_1} \cdot \prod_{r=2}^{p} \left((D(Bm_r+C))^{d_c}\right)^{(c_{\ell_r}-c_{\ell_1})/d_c} \\ &\quad \cdot \det\left(Q_{n,S,\{b_r\},\{c_r\},\{a_r\}}\right), \end{aligned} \tag{6.27}$$

where

$$\mathcal{Q}_{n,S,\{b_r\},\{c_r\},\{a_r\}} \equiv \mathcal{Q}_{n,S,\{b_1,\ldots,b_n\},\{c_1,\ldots,c_n\},\{a_r\}} \tag{6.28}$$

is defined to be the $n \times n$ matrix whose i-th row is

$$1, \left((D(Bm_r + C))^{d_b}\right)^{(b_2-b_1)/d_b}, \ldots, \left((D(Bm_r + C))^{d_b}\right)^{(b_n-b_1)/d_b},$$

if $i = \ell_r \in S$, and

$$a_i, a_{i+1}, \ldots, a_{i+n-1}, \quad \text{if } i \notin S. \tag{6.29}$$

Letting $\rho \in \mathcal{S}_p$ act on $\{m_1, m_2, \ldots, m_p\}$ by $m_r \to m_{\rho(r)}$, we observe that

$$\rho\left(\det\left(\mathcal{Q}_{n,S,\{b_r\},\{c_r\},\{a_r\}}\right)\right) = \operatorname{sign}(\rho) \cdot \det\left(\mathcal{Q}_{n,S,\{b_r\},\{c_r\},\{a_r\}}\right), \tag{6.30}$$

since permuting $\{m_1, m_2, \ldots, m_p\}$ by ρ just permutes the rows of $\mathcal{Q}_{n,S,\{b_r\},\{c_r\},\{a_r\}}$ corresponding to $i = \ell_r \in S$.

Furthermore, we have

$$\sum_{\rho\in\mathcal{S}_p} \operatorname{sign}(\rho) \prod_{r=2}^{p} \left((D(Bm_{\rho(r)} + C))^{d_c}\right)^{(c_{\ell_r}-c_{\ell_1})/d_c}$$
$$= \det\left(\left((D(Bm_r + C))^{d_c}\right)^{(c_{\ell_s}-c_{\ell_1})/d_c}\right)_{1\le r,s\le p}, \tag{6.31}$$

where in the left-hand-side of (6.31) we used the definition of determinant in which a different row element is selected from each column. Thus, the determinant in (6.31) is skew-symmetric in $\{m_1, m_2, \ldots, m_p\}$.

Finally, by symmetry, we have

$$\rho\left(\prod_{r=1}^{p}(D(Bm_r + C))^{c_{\ell_1}+b_1}\right) = \prod_{r=1}^{p}(D(Bm_r + C))^{c_{\ell_1}+b_1}. \tag{6.32}$$

Since the left-hand-side of (6.26) is independent of $\rho \in \mathcal{S}_p$, it now follows from (6.27)–(6.32), summing both sides of (6.26) over $\rho \in \mathcal{S}_p$, and interchange of summation that we have the following lemma.

Lemma 6.12. *Let $M_{n,S,\{b_r\},\{c_r\},\{a_r\}}$ be determined by Definition 6.2 and $\mathcal{Q}_{n,S,\{b_r\},\{c_r\},\{a_r\}}$ by (6.29). Let d_b and d_c be given by Definition 6.3. Let $n = 1, 2, \ldots$ and $p = 1, 2, \ldots, n$. Then,*

$$\det\left(M_{n,S,\{b_r\},\{c_r\},\{a_r\}}\right) = \frac{1}{p!}(-A)^{-p} q^{p(G-C)} \sum_{\substack{y_1,\ldots,y_p\ge1\\ m_1,\ldots,m_p\ge1}} q^{B(m_1y_1+\cdots+m_py_p)}$$
$$\cdot (-A)^{y_1+\cdots+y_p} E^{m_1+\cdots+m_p} q^{(F-B)(m_1+\cdots+m_p)} q^{C(y_1+\cdots+y_p)}$$

$$\cdot \prod_{r=1}^{p} (D(Bm_r + C))^{c_{\ell_1}+b_1} \cdot \det\left(Q_{n,S,\{b_r\},\{c_r\},\{a_r\}}\right)$$
$$\cdot \det\left(\left((D(Bm_r + C))^{d_c}\right)^{(c_{\ell_s}-c_{\ell_1})/d_c}\right)_{1\le r,s\le p}. \tag{6.33}$$

We next rewrite (6.33) by appealing to the following symmetrization lemma.

Lemma 6.13. *Let $F(y_1, \ldots, y_p; m_1, \ldots, m_p)$ be symmetric in $\{y_1, \ldots, y_p\}$, and in $\{m_1, \ldots, m_p\}$. Let $G(m_1, \ldots, m_p)$ be symmetric in $\{m_1, \ldots, m_p\}$. Furthermore, assume that F equals 0 if any of $\{m_1, \ldots, m_p\}$ are equal. Let $n = 1, 2, \ldots$ and $p = 1, 2, \ldots, n$. Then,*

$$\sum_{\substack{y_1,\ldots,y_p\ge 1\\ m_1,\ldots,m_p\ge 1}} F(y_1, \ldots, y_p; m_1, \ldots, m_p)G(m_1y_1, \ldots, m_py_p)$$
$$= p! \sum_{\substack{y_1,\ldots,y_p\ge 1\\ m_1>m_2>\cdots>m_p\ge 1}} F(y_1, \ldots, y_p; m_1, \ldots, m_p)G(m_1y_1, \ldots, m_py_p). \tag{6.34}$$

Proof: Since we can assume that the $\{m_1, \ldots, m_p\}$ are distinct, we write the left-hand-side of (6.34) as

$$\sum_{\substack{y_1,\ldots,y_p\ge 1\\ m_1>m_2>\cdots>m_p\ge 1}} \sum_{\rho\in S_p} F(y_1, \ldots, y_p; m_{\rho(1)}, \ldots, m_{\rho(p)})G(m_{\rho(1)}y_1, \ldots, m_{\rho(p)}y_p). \tag{6.35}$$

Next, by an interchange of summation, symmetry of F in $\{m_1, \ldots, m_p\}$, and relabelling the $\{y_1, \ldots, y_p\}$ as $\{y_{\rho(1)}, \ldots, y_{\rho(p)}\}$, we have

$$\sum_{\rho\in S_p} \sum_{\substack{y_{\rho(1)},\ldots,y_{\rho(p)}\ge 1\\ m_1>m_2>\cdots>m_p\ge 1}} F\big(y_{\rho(1)}, \ldots, y_{\rho(p)}; m_1, \ldots, m_p\big)G\big(m_{\rho(1)}y_{\rho(1)}, \ldots, m_{\rho(p)}y_{\rho(p)}\big). \tag{6.36}$$

The symmetry of F in $\{y_1, \ldots, y_p\}$, the symmetry of G in $\{m_1y_1, \ldots, m_py_p\}$, the fact that $y_{\rho(1)}, \ldots, y_{\rho(p)} \ge 1$ is equivalent to $y_1, \ldots, y_p \ge 1$, and an interchange of summation now gives

$$\sum_{\substack{y_1,\ldots,y_p\ge 1\\ m_1>m_2>\cdots>m_p\ge 1}} F(y_1, \ldots, y_p; m_1, \ldots, m_p)G(m_1y_1, \ldots, m_py_p) \sum_{\rho\in S_p} 1. \tag{6.37}$$

Since the innermost sum over S_p equals $p!$, we obtain the right-hand-side of (6.34). □

It follows immediately from Lemma 6.13 that Lemma 6.12 simplifies to the following lemma.

Lemma 6.14. *Let $M_{n,S,\{b_r\},\{c_r\},\{a_r\}}$ be determined by Definition 6.2 and $Q_{n,S,\{b_r\},\{c_r\},\{a_r\}}$ by (6.29). Let d_b and d_c be given by Definition 6.3. Let $n = 1, 2, \ldots$ and $p = 1, 2, \ldots, n$.*

Then,

$$\det\left(M_{n,S,\{b_r\},\{c_r\},\{a_r\}}\right) = (-A)^{-p} q^{p(G-C)} \sum_{\substack{y_1,\ldots,y_p\geq 1\\ m_1>m_2>\cdots>m_p\geq 1}} (-A)^{y_1+\cdots+y_p} E^{m_1+\cdots+m_p}$$

$$\cdot\, q^{(F-B)(m_1+\cdots+m_p)} q^{(Bm_1+C)y_1+\cdots+(Bm_p+C)y_p}$$

$$\cdot \prod_{r=1}^{p} (D(Bm_r+C))^{c_{\ell_1}+b_1} \cdot \det\left(Q_{n,S,\{b_r\},\{c_r\},\{a_r\}}\right)$$

$$\cdot \det\left(\left((D(Bm_r+C))^{d_c}\right)^{(c_{\ell_s}-c_{\ell_1})/d_c}\right)_{1\leq r,s\leq p}. \tag{6.38}$$

Proof: Take $G(m_1y_1,\ldots,m_py_p) := q^{B(m_1y_1+\cdots+m_py_p)}$, which is symmetric in $\{m_1y_1,\ldots,m_py_p\}$. Let $F(y_1,\ldots,y_p; m_1,\ldots,m_p)$ be the rest of the term in the double multiple sum on the right-hand-side of (6.33). The two factors of F which are determinants are each skew-symmetric in $\{m_1,\ldots,m_p\}$. Thus, their product is symmetric in $\{m_1,\ldots,m_p\}$, and vanishes if any of $\{m_1,\ldots,m_p\}$ are equal. All other factors of F are clearly symmetric in $\{y_1,\ldots,y_p\}$, or in $\{m_1,\ldots,m_p\}$. □

In order to finish the proof of Theorem 6.7 we utilize the Schur functions in Definition 6.5 and the Laplace expansion formula in Theorem 6.8 to rewrite the two determinants appearing in each term on the right-hand-side of (6.38).

By Eq. (6.12) of Definition 6.5, column operations, and the product formula for a Vandermonde determinant we have the following lemma.

Lemma 6.15. *Let $x_1,\ldots,x_p$ be indeterminant, and $1\leq\mu_1<\mu_2<\cdots<\mu_p$ be positive integers. Let $p=1,2,\ldots$. Then,*

$$\det\left(x_i^{\mu_j-\mu_1}\right)_{1\leq i,j\leq p} = (-1)^{\binom{p}{2}} \prod_{1\leq r<s\leq p} (x_r-x_s)\cdot s_\lambda(x_1,\ldots,x_p), \tag{6.39}$$

where $s_\lambda(x_1,\ldots,x_p)$ is the Schur function in (6.12), *with the partition $\lambda=(\lambda_1\geq\lambda_2\geq\cdots\geq\lambda_p\geq 0)$ given by*

$$\lambda_i = \mu_{p-i+1}-\mu_1+i-p, \quad \textit{for } i=1,2,\ldots,p. \tag{6.40}$$

Proof: By column operations and a trivial simplification we first have

$$\det\left(x_i^{\mu_j-\mu_1}\right)_{1\leq i,j\leq p} = (-1)^{\binom{p}{2}} \det\left(x_i^{\mu_{p-j+1}-\mu_1}\right)_{1\leq i,j\leq p}$$
$$= (-1)^{\binom{p}{2}} \det\left(x_i^{(\mu_{p-j+1}-\mu_1+j-p)+p-j}\right)_{1\leq i,j\leq p}. \tag{6.41}$$

Next, appeal to (6.12) and the product formula for a Vandermonde determinant. Finally, since the μ_i are strictly increasing, it is clear that the λ_i in (6.40) determine the partition $\lambda =$

$(\lambda_1 \geq \lambda_2 \geq \cdots \geq \lambda_p \geq 0)$. Just note that $\lambda_p = 0$ and $\lambda_i - \lambda_{i+1} = \mu_{p-i+1} - \mu_{p-i} - 1 \geq 0$, if $i = 1, 2, \ldots, p-1$. □

We now use Lemma 6.15 to rewrite the second determinant on the right-hand-side of (6.38). We specialize x_r and μ_r as follows:

$$x_r := (D(Bm_r + C))^{d_c}, \quad \text{for } r = 1, 2, \ldots, p, \tag{6.42}$$

$$\mu_r := c_{\ell_r}/d_c, \quad \text{for } r = 1, 2, \ldots, p. \tag{6.43}$$

Since ℓ_r and c_r are both strictly increasing and d_c is constant, then the $\mu_r = c_{\ell_r}/d_c$ are also strictly increasing. We obtain

$$\begin{aligned} &\det\left(\left((D(Bm_r + C))^{d_c}\right)^{(c_{\ell_s} - c_{\ell_1})/d_c}\right)_{1 \leq r,s \leq p} \\ &= (-1)^{\binom{p}{2}} \prod_{1 \leq r < s \leq p} \left((D(Bm_r + C))^{d_c} - (D(Bm_s + C))^{d_c}\right) \\ &\quad \cdot s_\lambda\left((D(Bm_1 + C))^{d_c}, \ldots, (D(Bm_p + C))^{d_c}\right), \end{aligned} \tag{6.44}$$

where λ_i is given by (6.13).

We next use Theorem 6.8 to expand the first determinant on the right-hand-side of (6.38) into the following sum.

$$\det\left(Q_{n,S,\{b_r\},\{c_r\},\{a_r\}}\right) = \sum_{\substack{\emptyset \subset T \subseteq I_n \\ \|T\| = p}} \det(C_{p,S,T}) \det\left(D_{n-p,S^c,T^c}\right) \cdot (-1)^{\Sigma(S) + \Sigma(T)}, \tag{6.45}$$

where S, T, S^c, T^c, $\Sigma(S)$, and $\Sigma(T)$ be given by (6.1)–(6.3), $Q_{n,S,\{b_r\},\{c_r\},\{a_r\}}$ is defined by (6.29), the $(n-p) \times (n-p)$ matrix D_{n-p,S^c,T^c} is determined by (6.11), and the $p \times p$ matrix $C_{p,S,T}$ is defined by

$$C_{p,S,T} := \left[\left((D(Bm_r + C))^{d_b}\right)^{(b_{j_s} - b_1)/d_b}\right]_{1 \leq r,s \leq p}. \tag{6.46}$$

By factoring $((D(Bm_r + C))^{d_b})^{(b_{j_1} - b_1)/d_b}$ out of the r-th row of $C_{p,S,T}$ it is immediate that

$$\det(C_{p,S,T}) = \prod_{r=1}^{p} \left((D(Bm_r + C))^{d_b}\right)^{(b_{j_1} - b_1)/d_b} \tag{6.47a}$$

$$\det\left(\left((D(Bm_r + C))^{d_b}\right)^{(b_{j_s} - b_{j_1})/d_b}\right)_{1 \leq r,s \leq p}. \tag{6.47b}$$

We now use Lemma 6.15 to rewrite the $p \times p$ determinant in (6.47b). We specialize x_r and μ_r as follows:

$$x_r := (D(Bm_r + C))^{d_b}, \quad \text{for } r = 1, 2, \ldots, p, \tag{6.48}$$

$$\mu_r := b_{j_r}/d_b, \quad \text{for } r = 1, 2, \ldots, p. \tag{6.49}$$

Since j_r and b_r are both strictly increasing and d_b is constant, then the $\mu_r = b_{j_r}/d_b$ are also strictly increasing. We obtain

$$\begin{aligned}
&\det\Big(\big((D(Bm_r+C))^{d_b}\big)^{(b_{j_s}-b_{j_1})/d_b}\Big)_{1\le r,s\le p} \\
&= (-1)^{\binom{p}{2}} \prod_{1\le r<s\le p} \big((D(Bm_r+C))^{d_b}-(D(Bm_s+C))^{d_b}\big) \\
&\cdot s_\nu\big((D(Bm_1+C))^{d_b},\ldots,(D(Bm_p+C))^{d_b}\big), \qquad (6.50)
\end{aligned}$$

where ν_i is given by (6.14).

By combining (6.45)–(6.50) we have the following lemma.

Lemma 6.16. *Let $Q_{n,S,\{b_r\},\{c_r\},\{a_r\}}$ be defined by (6.29). Let S, T, S^c, T^c, $\Sigma(S)$, $\Sigma(T)$ be given by (6.1)–(6.3), and let d_b and d_c be given by Definition 6.3. Let the $(n-p)\times(n-p)$ matrix D_{n-p,S^c,T^c} be determined by (6.11). Let the Schur function s_ν be defined by (6.12), with the partition $\nu = (\nu_1 \ge \nu_2 \ge \cdots \ge \nu_p \ge 0)$ given by (6.14). Then,*

$$\begin{aligned}
\det\big(Q_{n,S,\{b_r\},\{c_r\},\{a_r\}}\big) &= (-1)^{\binom{p}{2}} \prod_{1\le r<s\le p} \big((D(Bm_r+C))^{d_b}-(D(Bm_s+C))^{d_b}\big) \\
&\cdot \sum_{\substack{\emptyset\subset T\subseteq I_n \\ \|T\|=p}} (-1)^{\Sigma(S)+\Sigma(T)} \cdot \det\big(D_{n-p,S^c,T^c}\big)\cdot\left(\prod_{r=1}^{p} D(Bm_r+C)\right)^{b_{j_1}-b_1} \\
&\cdot s_\nu\big((D(Bm_1+C))^{d_b},\ldots,(D(Bm_p+C))^{d_b}\big). \qquad (6.51)
\end{aligned}$$

The proof of Theorem 6.7 is completed by substituting (6.44) and (6.51) into the right-hand-side of (6.38) and then simplifying.

The derivation of the Schur function form of our χ_n determinant identities in Section 7 requires a slight modification of Theorem 6.7, which arises from replacing $a_i, a_{i+1}, \ldots, a_{i+n-1}$ in (6.7) by $a_i, a_{i+1}, \ldots, a_{i+n-2}, a_{i+n}$. We first need the following two definitions.

Definition 6.17 (Second Lambert series matrix). Let $b_1, b_2, \ldots, b_n$ and $c_1, c_2, \ldots, c_n$ be indeterminant, and let $n = 1, 2, \ldots$. Furthermore, assume that $b_1 < b_2 < \cdots < b_n$ and $c_1 < c_2 < \cdots < c_n$. Take $\{a_r : r = 1, 2, \ldots\}$ to be an arbitrary sequence. Let S and S^c be the subsets of I_n in (6.1), with $p = 1, 2, \ldots, n$. Let $L_u(r; A, \ldots, G)$ be the Lambert series in Definition 6.1. Then,

$$\bar{M}_{n,S,\{b_r\},\{c_r\},\{a_r\}} \equiv \bar{M}_{n,S,\{b_1,\ldots,b_n\},\{c_1,\ldots,c_n\},\{a_r\}} \qquad (6.52)$$

is defined to be the $n\times n$ matrix whose i-th row is

$$L_{c_i+b_1}(m_\mu; A,\ldots,G),\ L_{c_i+b_2}(m_\mu; A,\ldots,G),\ \ldots,\ L_{c_i+b_n}(m_\mu; A,\ldots,G),$$

if $i = \ell_\mu \in S$, and

$$a_i, a_{i+1}, \ldots, a_{i+n-2}, a_{i+n}, \quad \text{if } i \notin S. \qquad (6.53)$$

Definition 6.18 (Second Laplace expansion formula determinant). Let $\{a_r : r = 1, 2, \ldots\}$ be an arbitrary sequence. Let the sets S^c and T^c be given by (6.1) and (6.2), and let $p = 1, 2, \ldots, n$. Then,

$$\det\left(\bar{D}_{n-p,S^c,T^c}\right) \tag{6.54}$$

is the determinant of the $(n-p) \times (n-p)$ matrix

$$\bar{D}_{n-p,S^c,T^c} := \left[a_{(\ell_{p+r}+j_{p+s}-1+\chi(j_{p+s}=n))}\right]_{1\le r,s\le n-p}, \tag{6.55}$$

where

$$\chi(A) := 1, \text{ if } A \text{ is true, and } 0, \text{ otherwise.} \tag{6.56}$$

The same computations that established Theorem 6.7 now give the following analogous expansion.

Theorem 6.19 (*Expansion of second Lambert series determinant*). *Let $L_u(r; A, \ldots, G)$ be the Lambert series in Definition 6.1. Let the $n \times n$ matrix $\bar{M}_{n,S,\{b_r\},\{c_r\},\{a_r\}}$ of Lambert series be given by Definition 6.17, and take $n = 1, 2, \ldots$ and $p = 1, 2, \ldots, n$. Let A, B, C, D, E, F, G, and $b_1, b_2, \ldots, b_n$ and $c_1, c_2, \ldots, c_n$ be indeterminant, and assume that $0 < |q| < 1$. Let S, T, S^c, T^c, $\Sigma(S)$, and $\Sigma(T)$ be given by (6.1)–(6.3). Take $\{a_r : r = 1, 2, \ldots\}$ to be an arbitrary sequence. Let d_b and d_c be given by Definition 6.3 and s_λ and s_ν be the Schur functions in Definition 6.5 with partitions λ and ν as in Definition 6.6. Finally, take the $(n-p) \times (n-p)$ matrix $\bar{D}_{n-p,S^c,T^c}$ in Definition 6.18. We then have the expansion formula*

$$\begin{aligned}
&\det\left(\bar{M}_{n,S,\{b_r\},\{c_r\},\{a_r\}}\right) \\
&\quad = (-A)^{-p} q^{p(G-C)} \sum_{\substack{y_1,\ldots,y_p\ge 1\\ m_1>m_2>\cdots>m_p\ge 1}} (-A)^{y_1+\cdots+y_p} E^{m_1+\cdots+m_p} \\
&\quad \cdot q^{(F-B)(m_1+\cdots+m_p)} q^{(Bm_1+C)y_1+\cdots+(Bm_p+C)y_p} \\
&\quad \cdot \prod_{1\le r<s\le p} \left((D(Bm_r+C))^{d_c} - (D(Bm_s+C))^{d_c}\right) \\
&\quad \cdot \prod_{1\le r<s\le p} \left((D(Bm_r+C))^{d_b} - (D(Bm_s+C))^{d_b}\right) \\
&\quad \cdot s_\lambda\left((D(Bm_1+C))^{d_c}, \ldots, (D(Bm_p+C))^{d_c}\right) \\
&\quad \cdot \sum_{\substack{\emptyset\subset T\subseteq I_n\\ \|T\|=p}} (-1)^{\Sigma(S)+\Sigma(T)} \cdot \det\left(\bar{D}_{n-p,S^c,T^c}\right) \cdot \left(\prod_{r=1}^{p} D(Bm_r+C)\right)^{c_{\ell_1}+b_{j_1}} \\
&\quad \cdot s_\nu\left((D(Bm_1+C))^{d_b}, \ldots, (D(Bm_p+C))^{d_b}\right).
\end{aligned} \tag{6.57}$$

7. The Schur function form of sums of squares identities

In this section we first apply the key determinant expansion formulas in Theorems 6.7 and 6.17 to most of the main identities in Section 5 to obtain the Schur function form of our infinite families of sums of squares and related identities. These include the two Kac–Wakimoto [120, p. 452] conjectured identities involving representing a positive integer by sums of $4n^2$ or $4n(n+1)$ triangular numbers, respectively. In addition, we obtain in Corollary 7.10 our analog of these Kac–Wakimoto identities which involves representing a positive integer by sums of $2n$ squares and $(2n)^2$ triangular numbers. Motivated by the analysis in [41, 42], we next use Jacobi's transformation of the theta functions ϑ_4 and ϑ_2 to derive a direct connection between our identities involving $4n^2$ or $4n(n+1)$ squares, and the identities involving $4n^2$ or $4n(n+1)$ triangular numbers. In a different direction we also apply the classical techniques in [90, pp. 96–101] and [28, pp. 288–305] to several of the theorems in this section to obtain the corresponding infinite families of lattice sum transformations.

We begin with the Schur function form of the $4n^2$ squares identity in the following theorem.

Theorem 7.1. *Let $n = 1, 2, 3, \ldots$. Then*

$$\vartheta_3(0,-q)^{4n^2} = 1 + \sum_{p=1}^{n} (-1)^p 2^{2n^2+n} \prod_{r=1}^{2n-1} (r!)^{-1} \sum_{\substack{y_1,\ldots,y_p\geq 1 \\ m_1>m_2>\cdots>m_p\geq 1}} (-1)^{y_1+\cdots+y_p}$$
$$\cdot (-1)^{m_1+\cdots+m_p} q^{m_1y_1+\cdots+m_py_p} \prod_{1\leq r<s\leq p} (m_r^2 - m_s^2)^2$$
$$\cdot (m_1m_2\ldots m_p) \sum_{\substack{\emptyset\subset S,T\subseteq I_n \\ \|S\|=\|T\|=p}} (-1)^{\Sigma(S)+\Sigma(T)} \cdot \det\left(D_{n-p,S^c,T^c}\right)$$
$$\cdot (m_1m_2\ldots m_p)^{2\ell_1+2j_1-4} s_\lambda\left(m_1^2,\ldots,m_p^2\right) s_\nu\left(m_1^2,\ldots,m_p^2\right), \tag{7.1}$$

where $\vartheta_3(0,-q)$ is determined by (1.1), *the sets S, S^c, T, T^c are given by* (6.1)–(6.2), *$\Sigma(S)$ and $\Sigma(T)$ by* (6.3), *the $(n-p)\times(n-p)$ matrix*

$$D_{n-p,S^c,T^c} := \left[c_{(\ell_{p+r}+j_{p+s}-1)}\right]_{1\leq r,s\leq n-p}, \tag{7.2}$$

where the c_i are defined by (5.23), *with the B_{2i} in* (2.60), *and s_λ and s_ν are the Schur functions in* (6.12), *with the partitions λ and ν given by*

$$\lambda_i := \ell_{p-i+1} - \ell_1 + i - p \quad \text{and} \quad \nu_i := j_{p-i+1} - j_1 + i - p, \quad \text{for } i = 1, 2, \ldots, p, \tag{7.3}$$

where the ℓ_i and j_i are elements of the sets S and T, respectively.

Proof: We apply Theorem 6.7 to Theorem 5.4. The Lambert series U_{2i-1} in (5.30), determined by (5.22), is

$$-L_{2i-1}(r; 1, 1, 0, 1, -1, 1, 0), \tag{7.4}$$

as defined in (6.4). Factor -1 out of each row for $i \in S$ of the matrix $M_{n,S}$ in (5.29). This gives $(-1)^p$. Next, apply the following case of Theorem 6.7 to the resulting $\det(M_{n,S})$ in the p-th term of Eq. (5.29) in Theorem 5.4:

$$A = 1, \quad B = 1, \quad C = 0, \quad D = 1, \quad E = -1, \quad F = 1, \quad G = 0, \tag{7.5}$$

$$a_i = (-1)^{i-1}\frac{(2^{2i}-1)}{4i}\cdot|B_{2i}|, \quad \text{for } i = 1, 2, 3, \ldots, \tag{7.6}$$

$$c_i = 2i - 1 := 2\ell_\mu - 1, \tag{7.7}$$

$$b_i = 2(i-1), \quad \text{for } i = 1, 2, \ldots, n, \tag{7.8}$$

$$d_b = d_c = 2, \tag{7.9}$$

with B_{2i} the Bernoulli numbers in (2.60).

It is immediate that

$$\begin{gathered} G - C = 0, \quad F - B = 0, \quad Bm_r + C = m_r, \\ (D(Bm_r + C))^{d_b} = (D(Bm_r + C))^{d_c} = m_r^2. \end{gathered} \tag{7.10}$$

The λ_i and ν_i are given by (7.3). Note that ℓ_{p-i+1} and j_{p-i+1} are substituted for i in (7.7) and (7.8), respectively, before computing the λ_i and ν_i in (6.13) and (6.14).

Interchanging the inner sum in (5.29) with the double multiple sum over $y_1, \ldots, y_p \geq 1$ and $m_1 > m_2 > \cdots > m_p \geq 1$ in (6.15), multiplying and dividing by $(m_1 m_2 \ldots m_p)$, and simplifying, now yields (7.1). □

The Schur function form of the $4n(n+1)$ squares identity is given by the following theorem.

Theorem 7.2. *Let $n = 1, 2, 3, \ldots$. Then*

$$\begin{aligned} \vartheta_3(0, -q)^{4n(n+1)} = 1 + \sum_{p=1}^{n} (-1)^{n-p} 2^{2n^2+3n} \prod_{r=1}^{2n} (r!)^{-1} \sum_{\substack{y_1,\ldots,y_p \geq 1 \\ m_1 > m_2 > \cdots > m_p \geq 1}} (-1)^{m_1+\cdots+m_p} \\ \cdot q^{m_1 y_1 + \cdots + m_p y_p} (m_1 m_2 \ldots m_p)^3 \prod_{1 \leq r < s \leq p} (m_r^2 - m_s^2)^2 \\ \cdot \sum_{\substack{\emptyset \subset S, T \subseteq I_n \\ \|S\| = \|T\| = p}} (-1)^{\Sigma(S)+\Sigma(T)} \cdot \det(D_{n-p, S^c, T^c}) \\ \cdot (m_1 m_2 \ldots m_p)^{2\ell_1 + 2j_1 - 4} s_\lambda(m_1^2, \ldots, m_p^2)\, s_\nu(m_1^2, \ldots, m_p^2), \end{aligned} \tag{7.11}$$

where the same assumptions hold as in Theorem 7.1, *except that the* $(n-p)\times(n-p)$ *matrix*

$$D_{n-p,S^c,T^c} := \left[a_{(\ell_{p+r}+j_{p+s}-1)}\right]_{1\le r,s\le n-p}, \tag{7.12}$$

where the a_i *are defined by* (5.40), *with the* B_{2i+2} *in* (2.60).

Proof: We apply Theorem 6.7 to Theorem 5.6. The Lambert series G_{2i+1} in (5.47), determined by (5.39), is

$$L_{2i+1}(r; -1, 1, 0, 1, -1, 1, 0), \tag{7.13}$$

as defined in (6.4). Apply the following case of Theorem 6.7 to the $\det(M_{n,S})$ in the p-th term of Eq. (5.46) in Theorem 5.6:

$$A=-1,\quad B=1,\quad C=0,\quad D=1,\quad E=-1,\quad F=1,\quad G=0, \tag{7.14}$$

$$a_i=(-1)^i\frac{(2^{2i+2}-1)}{4(i+1)}\cdot|B_{2i+2}|,\quad \text{for } i=1,2,3,\ldots, \tag{7.15}$$

$$c_i=2i+1:=2\ell_\mu+1, \tag{7.16}$$

$$b_i=2(i-1),\quad \text{for } i=1,2,\ldots,n, \tag{7.17}$$

$$d_b=d_c=2, \tag{7.18}$$

with B_{2i+2} the Bernoulli numbers in (2.60).

It is immediate that (7.10) holds. The λ_i and ν_i are given by (7.3). Note that ℓ_{p-i+1} and j_{p-i+1} are substituted for i in (7.16) and (7.17), respectively, before computing the λ_i and ν_i in (6.13) and (6.14).

Interchanging the inner sum in (5.46) with the double multiple sum over $y_1,\ldots,y_p\ge 1$ and $m_1>m_2>\cdots>m_p\ge 1$ in (6.15), multiplying and dividing by $(m_1m_2\ldots m_p)^3$, and simplifying, now yields (7.11). □

We have in Section 8 written down the $n=3$ cases of Theorems 7.1 and 7.2 which yield explicit formulas for 36 and 48 squares, respectively. For reference, we have also written down the $n=2$ cases.

The next two theorems require the Euler numbers E_n defined by (2.61). We first have the following theorem.

Theorem 7.3. *Let* $n=1,2,3,\ldots$. *Then*

$$\vartheta_3(0,q)^{2n(n-1)}\vartheta_3(0,-q)^{2n^2}$$
$$=1+\sum_{p=1}^{n}(-1)^p 2^{2n}\prod_{r=1}^{n-1}(2r)!^{-2}\sum_{\substack{y_1,\ldots,y_p\ge 1\\ m_1>m_2>\cdots>m_p\ge 1(\text{odd})}}(-1)^{y_1+\cdots+y_p}$$

$$
\begin{aligned}
&\cdot (-1)^{\frac{1}{2}(p+m_1+\cdots+m_p)} q^{m_1 y_1+\cdots+m_p y_p} \prod_{1 \le r < s \le p} \left(m_r^2 - m_s^2\right)^2 \\
&\cdot \sum_{\substack{\emptyset \subset S, T \subseteq I_n \\ \|S\|=\|T\|=p}} (-1)^{\Sigma(S)+\Sigma(T)} \cdot \det\left(D_{n-p,S^c,T^c}\right) \\
&\cdot (m_1 m_2 \ldots m_p)^{2\ell_1+2j_1-4} s_\lambda\left(m_1^2, \ldots, m_p^2\right) s_\nu\left(m_1^2, \ldots, m_p^2\right),
\end{aligned}
\tag{7.19}
$$

where the same assumptions hold as in Theorem 7.1, *except that the* $(n-p) \times (n-p)$ *matrix*

$$D_{n-p,S^c,T^c} := \left[b_{(\ell_{p+r}+j_{p+s}-1)}\right]_{1 \le r,s \le n-p}, \tag{7.20}$$

where the b_i *are defined by* (5.57), *with the* E_{2i-2} *in* (2.61).

Proof: We apply Theorem 6.7 to Theorem 5.8. The Lambert series R_{2i-2} in (5.67), determined by (5.56), is

$$-L_{2i-2}(r; 1, 2, -1, 1, -1, 2, -1), \tag{7.21}$$

as defined in (6.4). Factor -1 out of each row for $i \in S$ of the matrix $M_{n,S}$ in (5.66). This gives $(-1)^p$. Next, apply the following case of Theorem 6.7 to the resulting $\det(M_{n,S})$ in the p-th term of Eq. (5.66) in Theorem 5.8:

$$A = 1, \quad B = 2, \quad C = -1, \quad D = 1, \quad E = -1, \quad F = 2, \quad G = -1, \tag{7.22}$$
$$a_i = (-1)^{i-1} \cdot \tfrac{1}{4} \cdot |E_{2i-2}|, \quad \text{for } i = 1, 2, 3, \ldots, \tag{7.23}$$
$$c_i = 2i - 2 := 2\ell_\mu - 2, \tag{7.24}$$
$$b_i = 2(i-1), \quad \text{for } i = 1, 2, \ldots, n, \tag{7.25}$$
$$d_b = d_c = 2, \tag{7.26}$$

with E_{2i-2} the Euler numbers in (2.61).

It is immediate that

$$
\begin{aligned}
&G - C = 0, \quad F - B = 0, \quad Bm_r + C = 2m_r - 1, \\
&(D(Bm_r + C))^{d_b} = (D(Bm_r + C))^{d_c} = (2m_r - 1)^2.
\end{aligned}
\tag{7.27}
$$

The λ_i and ν_i are given by (7.3). Note that ℓ_{p-i+1} and j_{p-i+1} are substituted for i in (7.24) and (7.25), respectively, before computing the λ_i and ν_i in (6.13) and (6.14).

Interchanging the inner sum in (5.66) with the double multiple sum over $y_1, \ldots, y_p \ge 1$ and $m_1 > m_2 > \cdots > m_p \ge 1$ in (6.15), relabelling $(2m_r - 1)$ as m_r (odd), and simplifying, now yields (7.19). □

We next have the following theorem.

Theorem 7.4. *Let $n = 1, 2, 3, \ldots$. Then*

$$\begin{aligned}
&\vartheta_3(0,q)^{2n(n+1)}\vartheta_3(0,-q)^{2n^2}\\
&= 1 + \sum_{p=1}^{n}(-1)^{n-p}2^{2n}\prod_{r=1}^{n}(2r-1)!^{-2} \sum_{\substack{y_1,\ldots,y_p\geq 1\\ m_1>m_2>\cdots>m_p\geq 1 (\text{odd})}} (-1)^{y_1+\cdots+y_p}\\
&\quad \cdot (-1)^{\frac{1}{2}(p+m_1+\cdots+m_p)} q^{m_1y_1+\cdots+m_py_p} \prod_{1\leq r<s\leq p}\left(m_r^2-m_s^2\right)^2\\
&\quad \cdot (m_1m_2\ldots m_p)^2 \sum_{\substack{\emptyset\subset S,T\subseteq I_n\\ \|S\|=\|T\|=p}} (-1)^{\Sigma(S)+\Sigma(T)} \cdot \det\left(D_{n-p,S^c,T^c}\right)\\
&\quad \cdot (m_1m_2\ldots m_p)^{2\ell_1+2j_1-4} s_\lambda\left(m_1^2,\ldots,m_p^2\right) s_\nu\left(m_1^2,\ldots,m_p^2\right),
\end{aligned} \tag{7.28}$$

where the same assumptions hold as in Theorem 7.1, *except that the* $(n-p)\times(n-p)$ *matrix*

$$D_{n-p,S^c,T^c} := \left[b_{(\ell_{p+r}+j_{p+s})}\right]_{1\leq r,s\leq n-p}, \tag{7.29}$$

where the b_i are determined by (5.57), *with the E_{2i-2} in* (2.61).

Proof: We apply Theorem 6.7 to Theorem 5.10. The Lambert series R_{2i} in (5.86), determined by (5.56), is

$$-L_{2i}(r; 1, 2, -1, 1, -1, 2, -1), \tag{7.30}$$

as defined in (6.4). Factor -1 out of each row for $i \in S$ of the matrix $M_{n,S}$ in (5.85). This gives $(-1)^p$. Next, apply the following case of Theorem 6.7 to the resulting $\det(M_{n,S})$ in the p-th term of Eq. (5.85) in Theorem 5.10:

$$\begin{aligned}
&A=1,\quad B=2,\quad C=-1,\quad D=1,\quad E=-1,\quad F=2,\quad G=-1, && (7.31)\\
&a_i = (-1)^i\cdot\tfrac{1}{4}\cdot|E_{2i}|,\quad \text{for } i=1,2,3,\ldots, && (7.32)\\
&c_i = 2i := 2\ell_\mu, && (7.33)\\
&b_i = 2(i-1),\quad \text{for } i=1,2,\ldots,n, && (7.34)\\
&d_b = d_c = 2, && (7.35)
\end{aligned}$$

with E_{2i} the Euler numbers in (2.61).

It is immediate that (7.27) holds. The λ_i and ν_i are given by (7.3). Note that ℓ_{p-i+1} and j_{p-i+1} are substituted for i in (7.33) and (7.34), respectively, before computing the λ_i and ν_i in (6.13) and (6.14).

Interchanging the inner sum in (5.85) with the double multiple sum over $y_1, \ldots, y_p \geq 1$ and $m_1 > m_2 > \cdots > m_p \geq 1$ in (6.15), relabelling $(2m_r - 1)$ as m_r (odd), multiplying and dividing by $(m_1m_2\ldots m_p)^2$, and simplifying, now yields (7.28). □

The next two results complete our proof of the Kac–Wakimoto conjectured identities for triangular numbers in [120, p. 452].

We first have the following theorem.

Theorem 7.5. *Let $\vartheta_2(0,q)$ be defined by* (1.2), *and let $n = 1, 2, 3, \ldots$. We then have*

$$\vartheta_2(0,q)^{4n^2} = 4^{n(n+1)} \prod_{r=1}^{2n-1} (r!)^{-1} \sum_{\substack{y_1,\ldots,y_n \geq 1(\text{odd}) \\ m_1>m_2>\cdots>m_n\geq 1(\text{odd})}} m_1 m_2 \ldots m_n \cdot q^{m_1 y_1+\cdots+m_n y_n} \prod_{1\leq r<s\leq n} \left(m_r^2 - m_s^2\right)^2, \tag{7.36}$$

and

$$\vartheta_2(0,q^{1/2})^{4n(n+1)} = 2^{n(4n+5)} \prod_{r=1}^{2n} (r!)^{-1} \sum_{\substack{y_1,\ldots,y_n \geq 1(\text{odd}) \\ m_1>m_2>\cdots>m_n\geq 1}} (m_1 m_2 \ldots m_n)^3 \cdot q^{m_1 y_1+\cdots+m_n y_n} \prod_{1\leq r<s\leq n} \left(m_r^2 - m_s^2\right)^2. \tag{7.37}$$

Proof: We apply Theorem 6.7 to Theorem 5.11. The Lambert series C_{2i-1} in (5.93), determined by (5.95), is

$$L_{2i-1}\left(r; -1, 4, -2, \tfrac{1}{2}, 1, 2, -1\right), \tag{7.38}$$

as defined in (6.4). Apply the following case of Theorem 6.7 to the single $n \times n$ determinant in the right-hand-side of Eq. (5.93) in Theorem 5.11:

$$A = -1, \quad B = 4, \quad C = -2, \quad D = \tfrac{1}{2}, \quad E = 1, \quad F = 2, \quad G = -1, \tag{7.39}$$

$$c_i = 2i - 1 := 2\mu - 1, \tag{7.40}$$

$$b_i = 2(i-1), \quad \text{for } i = 1, 2, \ldots, n, \tag{7.41}$$

$$d_b = d_c = 2. \tag{7.42}$$

It is immediate that

$$G - C = 1, \quad F - B = -2, \quad Bm_r + C = 4m_r - 2,$$
$$(D(Bm_r + C))^{d_b} = (D(Bm_r + C))^{d_c} = (2m_r - 1)^2. \tag{7.43}$$

In the above application of the determinant expansion of (6.15) we also have

$$p = n, \quad S = T = I_n, \quad \text{and} \quad \ell_\mu = j_\mu = \mu, \quad \text{for } \mu = 1, 2, \ldots, n. \tag{7.44}$$

Equations (7.40)–(7.42) and (7.44) imply that

$$\lambda_i = 0 \quad \text{and} \quad \nu_i = 0, \quad \text{for } i = 1, 2, \ldots, n. \tag{7.45}$$

Thus, we have

$$s_\lambda(x) = s_\nu(x) = \det(D_{0,\emptyset,\emptyset}) = 1. \tag{7.46}$$

Keeping in mind (7.39)–(7.46), noting that the inner sum in (6.15) consists of just one term corresponding to $T = I_n$, observing that

$$1 - 2m_r + (2m_r - 1)2y_r = (2m_r - 1)(2y_r - 1), \tag{7.47}$$

relabelling $(2m_r - 1)$ and $(2y_r - 1)$ as m_r (odd) and y_r (odd), respectively, and simplifying, now yields (7.36).

The proof of (7.37) is similar to that of (7.36). The Lambert series D_{2i+1} in (5.94), determined by (5.96), is

$$L_{2i+1}\left(r; -1, 2, 0, \tfrac{1}{2}, 1, 1, 0\right), \tag{7.48}$$

as defined in (6.4). Apply the following case of Theorem 6.7 to the single $n \times n$ determinant in the right-hand-side of Eq. (5.94) in Theorem 5.11:

$$A = -1, \quad B = 2, \quad C = 0, \quad D = \tfrac{1}{2}, \quad E = 1, \quad F = 1, \quad G = 0, \tag{7.49}$$

$$c_i = 2i + 1 := 2\mu + 1, \tag{7.50}$$

$$b_i = 2(i-1), \quad \text{for } i = 1, 2, \ldots, n, \tag{7.51}$$

$$d_b = d_c = 2. \tag{7.52}$$

It is immediate that

$$\begin{gathered} G - C = 0, \quad F - B = -1, \quad Bm_r + C = 2m_r, \\ (D(Bm_r + C))^{d_b} = (D(Bm_r + C))^{d_c} = m_r^2. \end{gathered} \tag{7.53}$$

Keeping in mind (7.44)–(7.46) and (7.49)–(7.53), noting that the inner sum in (6.15) consists of just one term corresponding to $T = I_n$, observing that $-m_r + 2m_r y_r = m_r(2y_r - 1)$, relabelling $(2y_r - 1)$ as y_r (odd), and simplifying, now yields (7.37). □

The same analysis involving (5.18) that took Theorem 5.11 into Corollary 5.12 transforms Theorem 7.5 into the following Corollary.

Corollary 7.6. *Let $\Delta(q)$ be defined by (5.17), and let $n = 1, 2, 3, \ldots$. We then have*

$$\Delta(q^2)^{4n^2} = 4^{-n(n-1)} q^{-n^2} \prod_{r=1}^{2n-1} (r!)^{-1} \sum_{\substack{y_1, \ldots, y_n \geq 1 (\text{odd}) \\ m_1 > m_2 > \cdots > m_n \geq 1 (\text{odd})}} m_1 m_2 \ldots m_n \cdot q^{m_1 y_1 + \cdots + m_n y_n} \prod_{1 \leq r < s \leq n} \left(m_r^2 - m_s^2\right)^2, \tag{7.54}$$

and

$$\Delta(q)^{4n(n+1)} = 2^n q^{-n(n+1)/2} \prod_{r=1}^{2n} (r!)^{-1} \sum_{\substack{y_1,\ldots,y_n \geq 1 (\text{odd}) \\ m_1 > m_2 > \cdots > m_n \geq 1}} (m_1 m_2 \ldots m_n)^3 \cdot q^{m_1 y_1 + \cdots + m_n y_n} \prod_{1 \leq r < s \leq n} (m_r^2 - m_s^2)^2. \tag{7.55}$$

Remark. Taking the coefficient of q^N on both sides of (7.54) and (7.55), respectively, immediately gives the first and second Kac–Wakimoto conjectured identities for triangular numbers in [120, p. 452]. Our proof of Corollary 7.6 does not require inclusion/exclusion, and the analysis involving Schur functions is simpler than in (7.1), (7.11), (7.19), and (7.28). Zagier in [254] has also recently independently proven these two identities in [120, p. 452] by utilizing cusp forms. In addition, he proved the more general Conjecture 7.2 of Kac–Wakimoto [120, p. 451], and its rewritten $m = 2$ (first unproven) special case in [120, p. 451]. Recently, in [183], Ono utilized an elementary modular forms involution to transform Zagier's proof and/or formulation in [254] of Corollary 7.6 into elegant corresponding sums of squares formulas for $4n^2$ or $4n(n+1)$ squares, respectively. These formulas of Ono are different than those in Theorems 5.3–5.6, and Theorems 7.1 and 7.2. Information regarding affine superalgebras and Appell's generalized theta functions can be found in [121].

The $n = 1$ cases of (7.54) and (7.55) are the classical identities of Legendre [139], [21, Eqs. (ii) and (iii), p. 139] given by

Theorem 7.7 *(Legendre). Let $\Delta(q)$ be defined by* (5.17). *Then*

$$\Delta(q^2)^4 = \sum_{r=1}^{\infty} \frac{(2r-1)q^{2(r-1)}}{1 - q^{2(2r-1)}} = \sum_{\substack{y_1 \geq 1 (\text{odd}) \\ m_1 \geq 1 (\text{odd})}} m_1 \cdot q^{m_1 y_1 - 1}, \tag{7.56}$$

$$\Delta(q)^8 = \sum_{r=1}^{\infty} \frac{r^3 q^{r-1}}{1 - q^{2r}} = \sum_{\substack{y_1 \geq 1 (\text{odd}) \\ m_1 \geq 1}} m_1^3 \cdot q^{m_1 y_1 - 1}. \tag{7.57}$$

The Legendre identities involve the first sum over r in (7.56) and (7.57). The $n = 1$ case of the Kac–Wakimoto conjectured identities for triangular numbers in [120, p. 452] takes the coefficient of q^N on both sides of (7.56) and (7.57) while using the second double sum over y_1 and m_1.

We next consider the Schur function form of our Hankel determinant identities related to $\mathrm{cn}(u, k)$ and $\mathrm{dn}(u, k)$.

We begin with the following theorem related to $\mathrm{cn}(u, k)$.

Theorem 7.8. *Let* $\vartheta_2(0,q)$ *and* $\vartheta_3(0,q)$ *be defined by* (1.2) *and* (1.1), *respectively. Let* $n = 1, 2, 3, \ldots$. *We then have*

$$\vartheta_2(0,q)^{2n^2}\vartheta_3(0,q)^{2n(n-1)} = 4^n\prod_{r=1}^{n-1}(2r)!^{-2}\sum_{\substack{y_1,\ldots,y_n\geq 1(\text{odd})\\ m_1>m_2>\cdots>m_n\geq 1(\text{odd})}}(-1)^{\frac{1}{2}(-n+y_1+\cdots+y_n)} \cdot q^{\frac{1}{2}(m_1y_1+\cdots+m_ny_n)}\prod_{1\leq r<s\leq n}\left(m_r^2-m_s^2\right)^2. \tag{7.58}$$

Proof: We apply Theorem 6.7 to Theorem 5.13. The Lambert series T_{2i-2} in (5.109), determined by (5.110), is

$$L_{2i-2}\left(r;1,2,-1,1,1,1,-\tfrac{1}{2}\right), \tag{7.59}$$

as defined in (6.4). Apply the following case of Theorem 6.7 to the single $n\times n$ determinant in the right-hand-side of Eq. (5.109) in Theorem 5.13:

$$A=1,\quad B=2,\quad C=-1,\quad D=1,\quad E=1,\quad F=1,\quad G=-\tfrac{1}{2}, \tag{7.60}$$
$$c_i = 2i-2 := 2\mu-2, \tag{7.61}$$
$$b_i = 2(i-1),\quad \text{for } i=1,2,\ldots,n, \tag{7.62}$$
$$d_b = d_c = 2. \tag{7.63}$$

It is immediate that

$$\begin{aligned} &G-C=\tfrac{1}{2},\quad F-B=-1,\quad Bm_r+C=2m_r-1,\\ &(D(Bm_r+C))^{d_b}=(D(Bm_r+C))^{d_c}=(2m_r-1)^2. \end{aligned} \tag{7.64}$$

In the above application of the determinant expansion of (6.15) we also have (7.44)–(7.46). Keeping in mind (7.44)–(7.46) and (7.60)–(7.64), noting that the inner sum in (6.15) consists of just one term corresponding to $T = I_n$, observing that (7.47) holds, relabelling $(2m_r-1)$ and $(2y_r-1)$ as m_r (odd) and y_r (odd), respectively, and simplifying, now yields (7.58). □

Applying the relation (5.105) to the left-hand-side of (7.58) immediately implies that

$$\vartheta_2(0,q)^{2n}\vartheta_2(0,q^{1/2})^{4n(n-1)} = 4^{n^2}\prod_{r=1}^{n-1}(2r)!^{-2}\sum_{\substack{y_1,\ldots,y_n\geq 1(\text{odd})\\ m_1>m_2>\cdots>m_n\geq 1(\text{odd})}}(-1)^{\frac{1}{2}(-n+y_1+\cdots+y_n)} \cdot q^{\frac{1}{2}(m_1y_1+\cdots+m_ny_n)}\prod_{1\leq r<s\leq n}\left(m_r^2-m_s^2\right)^2. \tag{7.65}$$

Keeping in mind (7.65) and (5.18), the $n = 1$ case of Theorem 7.8 is equivalent to the identity in [21, Example (iv), p. 139]. Similarly, the $n = 2$ case of Theorem 7.8

leads to

$$q\Delta(q^2)^4 \cdot q\Delta(q)^8 = \tfrac{1}{64}\left[T_0T_4 - T_2^2\right] \tag{7.66a}$$

$$= 2^{-6} \sum_{\substack{y_1,y_2\geq 1 \\ m_1>m_2\geq 1(\mathrm{odd})}} (-1)^{y_1+y_2}\left(m_1^2 - m_2^2\right)^2 q^{m_1(y_1-\frac{1}{2})+m_2(y_2-\frac{1}{2})}, \tag{7.66b}$$

where T_{2i-2} is defined by (5.110).

Note that Theorem 7.7 and the product $q\Delta(q^2)^4 \cdot q\Delta(q)^8$ in (7.66a) implies that the 4-fold sum in (7.66b) is the product of the double sums in (7.56) and (7.57), each multiplied by q. That is, (7.66b) is the product of the sums in [21, Eqs. (ii) and (iii), p. 139]. The double sums in (7.56) and (7.57) have no alternating signs, while the sum in (7.66b) does.

We next have the following theorem.

Theorem 7.9. *Let* $\vartheta_2(0,q)$ *and* $\vartheta_3(0,q)$ *be defined by* (1.2) *and* (1.1), *respectively. Let* $n = 1, 2, 3, \ldots$. *We then have*

$$\vartheta_2(0,q)^{2n^2}\vartheta_3(0,q)^{2n(n+1)}$$

$$= 4^n \prod_{r=1}^{n}(2r-1)!^{-2} \sum_{\substack{y_1,\ldots,y_n\geq 1(\mathrm{odd}) \\ m_1>m_2>\cdots>m_n\geq 1(\mathrm{odd})}} (-1)^{\frac{1}{2}(-n+y_1+\cdots+y_n)}$$

$$\cdot\,(m_1m_2\ldots m_n)^2 q^{\frac{1}{2}(m_1y_1+\cdots+m_ny_n)} \prod_{1\leq r<s\leq n}\left(m_r^2 - m_s^2\right)^2. \tag{7.67}$$

Proof: We apply Theorem 6.7 to Theorem 5.14. The Lambert series T_{2i} in (5.117), determined by (5.110), is

$$L_{2i}\left(r; 1, 2, -1, 1, 1, 1, -\tfrac{1}{2}\right), \tag{7.68}$$

as defined in (6.4). Apply the following case of Theorem 6.7 to the single $n\times n$ determinant in the right-hand-side of Eq. (5.117) in Theorem 5.14:

$$A = 1,\quad B = 2,\quad C = -1,\quad D = 1,\quad E = 1,\quad F = 1,\quad G = -\tfrac{1}{2}, \tag{7.69}$$

$$c_i = 2i := 2\mu, \tag{7.70}$$

$$b_i = 2(i-1),\quad \text{for } i = 1, 2, \ldots, n, \tag{7.71}$$

$$d_b = d_c = 2. \tag{7.72}$$

It is immediate that (7.64) holds. In the above application of the determinant expansion of (6.15) we also have (7.44)–(7.46). Keeping in mind (7.44)–(7.46), (7.64), and (7.69)–(7.72), noting that the inner sum in (6.15) consists of just one term corresponding to $T = I_n$, observing that (7.47) holds, relabelling $(2m_r - 1)$ and $(2y_r - 1)$ as m_r (odd) and y_r (odd), respectively, and simplifying, now yields (7.67). □

Applying (5.18) and (5.105) to the left-hand-side of (7.67) immediately gives the Schur function form of Corollary 5.15 in the following corollary.

Corollary 7.10. *Let $\vartheta_3(0,q)$ and $\Delta(q)$ be defined by* (1.1) *and* (5.17), *respectively. Let* $n = 1, 2, 3, \ldots$. *We then have*

$$\vartheta_3(0,q)^{2n}\Delta(q)^{(2n)^2} = 4^{-n(n-1)}q^{-n^2/2}\prod_{r=1}^{n}(2r-1)!^{-2} \cdot \sum_{\substack{y_1,\ldots,y_n\geq 1(\text{odd})\\ m_1>m_2>\cdots>m_n\geq 1(\text{odd})}} (-1)^{\frac{1}{2}(-n+y_1+\cdots+y_n)}(m_1m_2\ldots m_n)^2 \cdot q^{\frac{1}{2}(m_1y_1+\cdots+m_ny_n)} \prod_{1\leq r<s\leq n}\left(m_r^2-m_s^2\right)^2. \tag{7.73}$$

We now consider the Schur function form of the theta function identities related to $\mathrm{dn}(u,k)$. We start with the following theorem.

Theorem 7.11. *Let $\vartheta_2(0,q)$ and $\vartheta_3(0,q)$ be defined by* (1.2) *and* (1.1), *respectively. Let* $n = 1, 2, 3, \ldots$. *We then have*

$$\vartheta_2(0,q)^{2n(n-1)}\vartheta_3(0,q)^{2n^2} = 4^{n^2}\prod_{r=1}^{n-1}(2r)!^{-2} \sum_{\substack{y_1,\ldots,y_n\geq 1(\text{odd})\\ m_1>m_2>\cdots>m_n\geq 1}} (-1)^{\frac{1}{2}(-n+y_1+\cdots+y_n)} \cdot q^{m_1y_1+\cdots+m_ny_n}\prod_{1\leq r<s\leq n}\left(m_r^2-m_s^2\right)^2 \tag{7.74a}$$

$$+4^{n^2-1}\prod_{r=1}^{n-1}(2r)!^{-2} \sum_{\substack{y_1,\ldots,y_{n-1}\geq 1(\text{odd})\\ m_1>m_2>\cdots>m_{n-1}\geq 1}} (-1)^{\frac{1}{2}(-n+1+y_1+\cdots+y_{n-1})} \cdot q^{m_1y_1+\cdots+m_{n-1}y_{n-1}}(m_1m_2\ldots m_{n-1})^4\prod_{1\leq r<s\leq n-1}\left(m_r^2-m_s^2\right)^2, \tag{7.74b}$$

where the term in (7.74b) *is defined to be* 1 *if* $n = 1$.

Proof: We apply Theorem 6.7 to Theorem 5.16. The Lambert series N_{2i-2} in (5.127a), determined by (5.128), is

$$L_{2i-2}\left(r; 1, 2, 0, \tfrac{1}{2}, 1, 1, 0\right), \tag{7.75}$$

as defined in (6.4). Apply the following case of Theorem 6.7 to the single $n\times n$ determinant in the right-hand-side of Eq. (5.127a) in Theorem 5.16:

$$A = 1,\quad B = 2,\quad C = 0,\quad D = \tfrac{1}{2},\quad E = 1,\quad F = 1,\quad G = 0, \tag{7.76}$$

$$c_i = 2i - 2 := 2\mu - 2, \tag{7.77}$$

$$b_i = 2(i-1), \quad \text{for } i = 1, 2, \ldots, n, \tag{7.78}$$
$$d_b = d_c = 2. \tag{7.79}$$

It is immediate that (7.53) holds. In the above application of the determinant expansion of (6.15) we also have (7.44)–(7.46). Keeping in mind (7.44)–(7.46), (7.53), and (7.76)–(7.79), noting that the inner sum in (6.15) consists of just one term corresponding to $T = I_n$, observing that $-m_r + 2m_r y_r = m_r(2y_r - 1)$, relabelling $(2y_r - 1)$ as y_r (odd), and simplifying, now yields the term in (7.74a).

The derivation of the term in (7.74b) from (5.127b) is the same as for (7.74a), except that $n \mapsto n-1$, $c_i \mapsto 2i+2$, and $c_{\ell_1} + b_{j_1} \mapsto 4$. □

The $n = 1$ case of Theorem 7.11 is equivalent to Jacobi's 2-squares identity in [21, Entry 8(i), p. 114]. The calculation is similar to that for the $n = 1$ case of Theorem 5.16. Here, just note that the $n = 1$ case of the sum in (7.74b) equals 1.

We next have the following theorem.

Theorem 7.12. *Let $\vartheta_2(0, q)$ and $\vartheta_3(0, q)$ be defined by* (1.2) *and* (1.1), *respectively. Let $n = 1, 2, 3, \ldots$. We then have*

$$\vartheta_2(0,q)^{2n(n+1)}\vartheta_3(0,q)^{2n^2} = 4^{n(n+1)} \prod_{r=1}^{n} (2r-1)!^{-2} \cdot \sum_{\substack{y_1,\ldots,y_n \geq 1 (\text{odd}) \\ m_1 > m_2 > \cdots > m_n \geq 1}} (-1)^{\frac{1}{2}(-n+y_1+\cdots+y_n)} (m_1 m_2 \ldots m_n)^2 \cdot q^{m_1 y_1 + \cdots + m_n y_n} \prod_{1 \leq r < s \leq n} \left(m_r^2 - m_s^2\right)^2. \tag{7.80}$$

Proof: We apply Theorem 6.7 to Theorem 5.17. The Lambert series N_{2i} in (5.135), determined by (5.128), is

$$L_{2i}\left(r; 1, 2, 0, \tfrac{1}{2}, 1, 1, 0\right), \tag{7.81}$$

as defined in (6.4). Apply the following case of Theorem 6.7 to the single $n \times n$ determinant in the right-hand-side of Eq. (5.135) in Theorem 5.17:

$$A = 1, \quad B = 2, \quad C = 0, \quad D = \tfrac{1}{2}, \quad E = 1, \quad F = 1, \quad G = 0, \tag{7.82}$$
$$c_i = 2i := 2\mu, \tag{7.83}$$
$$b_i = 2(i-1), \quad \text{for } i = 1, 2, \ldots, n, \tag{7.84}$$
$$d_b = d_c = 2. \tag{7.85}$$

It is immediate that (7.53) holds. In the above application of the determinant expansion of (6.15) we also have (7.44)–(7.46). Keeping in mind (7.44)–(7.46), (7.53), and (7.82)–(7.85), noting that the inner sum in (6.15) consists of just one term corresponding to $T = I_n$, observing that $-m_r + 2m_r y_r = m_r(2y_r - 1)$, relabelling $(2y_r - 1)$ as y_r (odd), and simplifying, now yields (7.80). □

Applying (5.18) and (5.105) to the left-hand-side of (7.80) immediately gives the Schur function form of Corollary 5.18 in the following corollary.

Corollary 7.13. *Let $\Delta(q)$ be defined by* (5.17). *Let $n = 1, 2, 3, \ldots$. We then have*

$$\Delta(q)^{(2n)^2}\Delta(q^2)^{2n} = q^{-n(n+1)/2}\prod_{r=1}^{n}(2r-1)!^{-2} \cdot \sum_{\substack{y_1,\ldots,y_n\geq 1(\text{odd})\\ m_1>m_2>\cdots>m_n\geq 1}} (-1)^{\frac{1}{2}(-n+y_1+\cdots+y_n)}(m_1m_2\ldots m_n)^2 \cdot q^{m_1y_1+\cdots+m_ny_n}\prod_{1\leq r<s\leq n}(m_r^2-m_s^2)^2. \tag{7.86}$$

We next survey the Schur function form of our χ_n determinant identities. We find it convenient to recall the definition of $\chi(A)$ given by

$$\chi(A) := 1, \text{ if } A \text{ is true, and } 0, \text{ otherwise.} \tag{7.87}$$

We first have the following theorem.

Theorem 7.14. *Let $n = 1, 2, 3, \ldots$. Then*

$$\vartheta_3(0,-q)^{4n^2}\left[1+24\sum_{r=1}^{\infty}\frac{rq^r}{1+q^r}\right] = 1+\sum_{p=1}^{n}(-1)^{p-1}2^{2n^2+n+1}\tfrac{3}{n(4n^2-1)}\prod_{r=1}^{2n-1}(r!)^{-1}\sum_{\substack{y_1,\ldots,y_p\geq 1\\ m_1>m_2>\cdots>m_p\geq 1}}(-1)^{y_1+\cdots+y_p} \cdot (-1)^{m_1+\cdots+m_p}q^{m_1y_1+\cdots+m_py_p}\prod_{1\leq r<s\leq p}(m_r^2-m_s^2)^2 \cdot (m_1m_2\ldots m_p)\sum_{\substack{\emptyset\subset S,T\subseteq I_n\\ \|S\|=\|T\|=p}}(-1)^{\Sigma(S)+\Sigma(T)}\cdot\det\left(\bar{D}_{n-p,S^c,T^c}\right) \cdot (m_1m_2\ldots m_p)^{2\ell_1+2j_1-4+2\chi(j_1=n)}s_\lambda(m_1^2,\ldots,m_p^2)\,s_\nu(m_1^2,\ldots,m_p^2), \tag{7.88}$$

where $\vartheta_3(0,-q)$ is determined by (1.1), *the sets S, S^c, T, T^c are given by* (6.1)–(6.2), *$\Sigma(S)$ and $\Sigma(T)$ by* (6.3), *and the $(n-p)\times(n-p)$ matrix*

$$\bar{D}_{n-p,S^c,T^c} := \left[c_{(\ell_{p+r}+j_{p+s}-1+\chi(j_{p+s}=n))}\right]_{1\leq r,s\leq n-p}, \tag{7.89}$$

where the c_i are determined by (5.23), with the B_{2i} in (2.60), and $\chi(A)$ defined by (7.87). Finally, s_λ and s_ν are the Schur functions in (6.12), with the partitions λ and ν given by

$$\lambda_i := \ell_{p-i+1} - \ell_1 + i - p, \quad \text{for } i = 1, 2, \ldots, p, \tag{7.90}$$

$$\nu_1 := 0, \quad \text{if } p = 1, \tag{7.91a}$$

$$\nu_1 := j_p - j_1 + 1 - p + \chi(j_p = n), \quad \text{if } p > 1, \tag{7.91b}$$

$$\nu_i := j_{p-i+1} - j_1 + i - p, \quad \text{if } 2 \leq i \leq p, \text{ and, } p > 1. \tag{7.91c}$$

Proof: We apply Theorem 6.19 to Theorem 5.24. We utilize the same specializations as in (7.4)–(7.9), except that

$$b_i = 2(i-1) + 2\chi(i = n), \quad \text{for } i = 1, 2, \ldots, n, \tag{7.92}$$

where $\chi(A)$ is defined by (7.87).

It is immediate that (7.10) holds. The λ_i and ν_i are given by (7.90)–(7.91). (Note that the formula for ν_i in (7.3) is transformed to (7.91a–c) by adding 1 to the largest part of ν if the largest element of the set T is n).

Interchanging the inner sum in (5.192) with the double multiple sum over $y_1, \ldots, y_p \geq 1$ and $m_1 > m_2 > \cdots > m_p \geq 1$ in (6.57), multiplying and dividing by $(m_1 m_2 \cdots m_p)$, and simplifying, now yields (7.88), for $n \geq 2$.

By (5.194), the $n = 1$ case of (7.88) is established by showing that this case of the multiple sum in (7.88) is U_3, as defined by (5.22). This equality follows termwise by using (6.5) to compute the $i = 2$ case of (7.4). □

We next have the following theorem.

Theorem 7.15. *Let $n = 1, 2, 3, \ldots$. Then*

$$\begin{aligned}
&\vartheta_3(0, -q)^{4n(n+1)} \left[1 + 24 \sum_{r=1}^{\infty} \frac{rq^r}{1+q^r}\right] \\
&= 1 + \sum_{p=1}^{n} (-1)^{n-p+1} 2^{2n^2+3n} \frac{3}{n(n+1)(2n+1)} \prod_{r=1}^{2n} (r!)^{-1} \sum_{\substack{y_1, \ldots, y_p \geq 1 \\ m_1 > m_2 > \cdots > m_p \geq 1}} (-1)^{m_1 + \cdots + m_p} \\
&\quad \cdot q^{m_1 y_1 + \cdots + m_p y_p} (m_1 m_2 \ldots m_p)^3 \prod_{1 \leq r < s \leq p} (m_r^2 - m_s^2)^2 \\
&\quad \cdot \sum_{\substack{\emptyset \subset S, T \subseteq I_n \\ \|S\| = \|T\| = p}} (-1)^{\Sigma(S) + \Sigma(T)} \cdot \det\left(\bar{D}_{n-p, S^c, T^c}\right) \\
&\quad \cdot (m_1 m_2 \ldots m_p)^{2\ell_1 + 2j_1 - 4 + 2\chi(j_1 = n)} s_\lambda(m_1^2, \ldots, m_p^2)\, s_\nu(m_1^2, \ldots, m_p^2),
\end{aligned} \tag{7.93}$$

where the same assumptions hold as in Theorem 7.14, except that the $(n-p) \times (n-p)$ matrix

$$\bar{D}_{n-p, S^c, T^c} := \left[a_{(\ell_{p+r} + j_{p+s} - 1 + \chi(j_{p+s} = n))}\right]_{1 \leq r, s \leq n-p}, \tag{7.94}$$

where the a_i are determined by (5.40), *with the* B_{2i+2} *in* (2.60), *and* $\chi(A)$ *defined by* (7.87).

Proof: We apply Theorem 6.19 to Theorem 5.26. We utilize the same specializations as in (7.13)–(7.18), except that b_i is given by (7.92).

It is immediate that (7.10) holds. The λ_i and ν_i are given by (7.90)–(7.91).

Interchanging the inner sum in (5.207) with the double multiple sum over $y_1, \ldots, y_p \geq 1$ and $m_1 > m_2 > \cdots > m_p \geq 1$ in (6.57), multiplying and dividing by $(m_1 m_2 \ldots m_p)^3$, and simplifying, now yields (7.93), for $n \geq 2$.

By (5.209), the $n = 1$ case of (7.93) is established by showing that this case of the multiple sum in (7.93) is G_5, as defined by (5.39). This equality follows termwise by using (6.5) to compute the $i = 2$ case of (7.13). □

The next two theorems require the Euler numbers E_n defined by (2.61). We first have the following theorem.

Theorem 7.16. *Let $n = 1, 2, 3, \ldots$. Then*

$$\vartheta_3(0,q)^{2n(n-1)}\vartheta_3(0,-q)^{2n^2} \cdot \left\{2n\left[2+24\sum_{r=1}^{\infty}\frac{2rq^{2r}}{1+q^{2r}}\right]-\left[1-24\sum_{r=1}^{\infty}\frac{(2r-1)q^{2r-1}}{1+q^{2r-1}}\right]\right\}$$
$$= (4n-1)+\sum_{p=1}^{n}(-1)^{p-1}2^{2n}\frac{3}{n(2n-1)}\prod_{r=1}^{n-1}(2r)!^{-2}\sum_{\substack{y_1,\ldots,y_p\geq 1\\ m_1>m_2>\cdots>m_p\geq 1(\text{odd})}}(-1)^{y_1+\cdots+y_p}$$
$$\cdot(-1)^{\frac{1}{2}(p+m_1+\cdots+m_p)}q^{m_1y_1+\cdots+m_py_p}\prod_{1\leq r<s\leq p}\left(m_r^2-m_s^2\right)^2$$
$$\cdot\sum_{\substack{\emptyset\subset S,T\subseteq I_n\\ \|S\|=\|T\|=p}}(-1)^{\Sigma(S)+\Sigma(T)}\cdot\det\left(\bar{D}_{n-p,S^c,T^c}\right)$$
$$\cdot(m_1m_2\ldots m_p)^{2\ell_1+2j_1-4+2\chi(j_1=n)}s_\lambda\left(m_1^2,\ldots,m_p^2\right)s_\nu\left(m_1^2,\ldots,m_p^2\right), \tag{7.95}$$

where the same assumptions hold as in Theorem 7.14, *except that the* $(n-p)\times(n-p)$ *matrix*

$$\bar{D}_{n-p,S^c,T^c} := \left[b_{(\ell_{p+r}+j_{p+s}-1+\chi(j_{p+s}=n))}\right]_{1\leq r,s\leq n-p}, \tag{7.96}$$

where the b_i are defined by (5.57), *with the* E_{2i-2} *in* (2.61), *and* $\chi(A)$ *defined by* (7.87).

Proof: We apply Theorem 6.19 to Theorem 5.28. We utilize the same specializations as in (7.21)–(7.26), except that b_i is given by (7.92).

It is immediate that (7.27) holds. The λ_i and ν_i are given by (7.90)–(7.91).

Interchanging the inner sum in (5.223) with the double multiple sum over $y_1, \ldots, y_p \geq 1$ and $m_1 > m_2 > \cdots > m_p \geq 1$ in (6.57), relabelling $(2m_r - 1)$ as m_r (odd), and simplifying, now yields (7.95), for $n \geq 2$.

By (5.225), the $n = 1$ case of (7.95) is established by showing that this case of the multiple sum in (7.95) is R_2, as defined by (5.56). This equality follows termwise by using (6.5) to compute the $i = 2$ case of (7.21). □

We next have the following theorem.

Theorem 7.17. *Let $n = 1, 2, 3, \ldots$. Then*

$$\vartheta_3(0,q)^{2n(n+1)}\vartheta_3(0,-q)^{2n^2}$$
$$\cdot\left\{2n\left[2+24\sum_{r=1}^{\infty}\frac{2rq^{2r}}{1+q^{2r}}\right]+\left[1-24\sum_{r=1}^{\infty}\frac{(2r-1)q^{2r-1}}{1+q^{2r-1}}\right]\right\}$$
$$= (4n+1)+\sum_{p=1}^{n}(-1)^{n-p+1}2^{2n}\frac{3}{n(2n+1)}\prod_{r=1}^{n}(2r-1)!^{-2}\sum_{\substack{y_1,\ldots,y_p\geq 1\\ m_1>m_2>\cdots>m_p\geq 1(\text{odd})}}(-1)^{y_1+\cdots+y_p}$$
$$\cdot(-1)^{\frac{1}{2}(p+m_1+\cdots+m_p)}q^{m_1y_1+\cdots+m_py_p}\prod_{1\leq r<s\leq p}(m_r^2-m_s^2)^2$$
$$\cdot(m_1m_2\ldots m_p)^2\sum_{\substack{\emptyset\subset S,T\subseteq I_n\\ \|S\|=\|T\|=p}}(-1)^{\Sigma(S)+\Sigma(T)}\cdot\det\left(\bar{D}_{n-p,S^c,T^c}\right)$$
$$\cdot(m_1m_2\ldots m_p)^{2\ell_1+2j_1-4+2\chi(j_1=n)}s_\lambda\left(m_1^2,\ldots,m_p^2\right)s_\nu\left(m_1^2,\ldots,m_p^2\right), \tag{7.97}$$

where the same assumptions hold as in Theorem 7.14, except that the $(n-p)\times(n-p)$ matrix

$$\bar{D}_{n-p,S^c,T^c} := \left[b_{(\ell_{p+r}+j_{p+s}+\chi(j_{p+s}=n))}\right]_{1\leq r,s\leq n-p}, \tag{7.98}$$

where the b_i are determined by (5.57), with the E_{2i-2} in (2.61), and $\chi(A)$ defined by (7.87).

Proof: We apply Theorem 6.19 to Theorem 5.30. We utilize the same specializations as in (7.30)–(7.35), except that b_i is given by (7.92).

It is immediate that (7.27) holds. The λ_i and ν_i are given by (7.90)–(7.91).

Interchanging the inner sum in (5.239) with the double multiple sum over $y_1, \ldots, y_p \geq 1$ and $m_1 > m_2 > \cdots > m_p \geq 1$ in (6.57), relabelling $(2m_r - 1)$ as m_r (odd), multiplying and dividing by $(m_1m_2\ldots m_p)^2$, and simplifying, now yields (7.97), for $n \geq 2$.

By (5.241), the $n = 1$ case of (7.97) is established by showing that this case of the multiple sum in (7.97) is R_4, as defined by (5.56). This equality follows termwise by using (6.5) to compute the $i = 2$ case of (7.30). □

The analog of Theorem 7.5 is given by the following theorem.

Theorem 7.18. *Let $\vartheta_2(0,q)$ be defined by* (1.2), *and let $n = 1, 2, 3, \ldots$. We then have*

$$\vartheta_2(0,q)^{4n^2}\left[1+24\sum_{r=1}^{\infty}\frac{rq^{2r}}{1+q^{2r}}\right]$$
$$= 4^{n(n+1)}\frac{3}{n(4n^2-1)}\prod_{r=1}^{2n-1}(r!)^{-1}\sum_{\substack{y_1,\ldots,y_n\geq 1(\text{odd})\\ m_1>m_2>\cdots>m_n\geq 1(\text{odd})}} m_1m_2\ldots m_n$$
$$\cdot\left(m_1^2+\cdots+m_n^2\right)q^{m_1y_1+\cdots+m_ny_n}\prod_{1\leq r<s\leq n}\left(m_r^2-m_s^2\right)^2, \tag{7.99}$$

and

$$\vartheta_2(0,q^{1/2})^{4n(n+1)}\left[1+24\sum_{r=1}^{\infty}\frac{rq^{r}}{1+q^{r}}\right]$$
$$= 2^{n(4n+5)}\frac{6}{n(n+1)(2n+1)}\prod_{r=1}^{2n}(r!)^{-1}\sum_{\substack{y_1,\ldots,y_n\geq 1(\text{odd})\\ m_1>m_2>\cdots>m_n\geq 1}} (m_1m_2\ldots m_n)^3$$
$$\cdot\left(m_1^2+\cdots+m_n^2\right)q^{m_1y_1+\cdots+m_ny_n}\prod_{1\leq r<s\leq n}\left(m_r^2-m_s^2\right)^2. \tag{7.100}$$

Proof: We apply Theorem 6.19 to Theorems 5.31 and 5.32. We take $n \geq 2$ and first consider (7.99). We utilize the same specializations as in (7.38)–(7.42), except that b_i is given by (7.92). It is immediate that (7.43) holds.

In the above application of the determinant expansion of (6.57) we also have (7.44). Equations (7.40), (7.42), (7.44), and (7.92) imply that

$$\lambda_i = 0, \quad \text{for } i = 1, 2, \ldots, n, \tag{7.101}$$
$$\nu_1 = 1, \quad \text{and} \quad \nu_i = 0, \quad \text{for } i = 2, 3, \ldots, n. \tag{7.102}$$

Thus, we have

$$s_\lambda(x) = \det(\bar{D}_{0,\emptyset,\emptyset}) = 1, \tag{7.103}$$
$$s_\nu(x_1,\ldots,x_n) = x_1+\cdots+x_n. \tag{7.104}$$

Keeping in mind (7.39), (7.40), (7.42), (7.43), (7.44), (7.92), (7.101)–(7.104), noting that the inner sum in (6.57) consists of just one term corresponding to $T = I_n$, observing (7.47), relabelling $(2m_r - 1)$ and $(2y_r - 1)$ as m_r (odd) and y_r (odd), respectively, and simplifying, now yields (7.99), for $n \geq 2$.

By Theorem 5.31, the $n = 1$ case of (7.99) is established by showing that this case of the multiple sum in (7.99) is C_3, as defined by (5.95). This equality follows termwise by using (6.5) to compute the $i = 2$ case of (7.38).

The proof of (7.100) is similar to that of (7.99). We take $n \geq 2$ and apply Theorem 6.19 to Theorem 5.32. We utilize the same specializations as in (7.48)–(7.52), except that b_i is given by (7.92). It is immediate that (7.53) holds.

Keeping in mind (7.44), (7.49), (7.50), (7.52), (7.53), (7.92), (7.101)–(7.104), noting that the inner sum in (6.57) consists of just one term corresponding to $T = I_n$, observing that $-m_r + 2m_r y_r = m_r(2y_r - 1)$, relabelling $(2y_r - 1)$ as y_r (odd), and simplifying, now yields (7.100), for $n \geq 2$.

By Theorem 5.32, the $n = 1$ case of (7.100) is established by showing that this case of the multiple sum in (7.100) is D_5, as defined by (5.96). This equality follows termwise by using (6.5) to compute the $i = 2$ case of (7.48). □

The same analysis involving (5.18) that took Theorem 5.11 into Corollary 5.12 transforms Theorem 7.18 into an analog of Corollary 7.6.

We next consider the Schur function form of our χ_n determinant identities related to $\mathrm{cn}(u, k)$, $\mathrm{sn}(u, k)\,\mathrm{dn}(u, k)$, $\mathrm{dn}(u, k)$, and $\mathrm{sn}(u, k)\,\mathrm{cn}(u, k)$.

We first have the following theorem.

Theorem 7.19. *Let $\vartheta_2(0, q)$ and $\vartheta_3(0, q)$ be defined by* (1.2) *and* (1.1), *respectively. Let $n = 1, 2, 3, \ldots$. We then have*

$$\vartheta_2(0,q)^{2n^2}\vartheta_3(0,q)^{2n(n-1)} \cdot \left\{2n\left[1+24\sum_{r=1}^{\infty}\frac{rq^r}{1+q^r}\right]+\left[1-24\sum_{r=1}^{\infty}\frac{(2r-1)q^{2r-1}}{1+q^{2r-1}}\right]\right\}$$
$$= 4^n \tfrac{3}{n(2n-1)}\prod_{r=1}^{n-1}(2r)!^{-2} \sum_{\substack{y_1,\ldots,y_n\geq 1(\mathrm{odd})\\ m_1>m_2>\cdots>m_n\geq 1(\mathrm{odd})}} (-1)^{\frac{1}{2}(-n+y_1+\cdots+y_n)} \cdot \left(m_1^2+\cdots+m_n^2\right) q^{\frac{1}{2}(m_1y_1+\cdots+m_ny_n)} \prod_{1\leq r<s\leq n}\left(m_r^2-m_s^2\right)^2. \tag{7.105}$$

Proof: We take $n \geq 2$ and apply Theorem 6.19 to Theorem 5.33. We utilize the same specializations as in (7.59)–(7.63), except that b_i is given by (7.92). It is immediate that (7.64) holds. In this application of the determinant expansion of (6.57) we also have (7.44), and (7.101)–(7.104).

Keeping in mind (7.44), (7.60), (7.61), (7.63), (7.64), (7.92), (7.101)–(7.104), noting that the inner sum in (6.57) consists of just one term corresponding to $T = I_n$, observing that (7.47) holds, relabelling $(2m_r - 1)$ and $(2y_r - 1)$ as m_r (odd) and y_r (odd), respectively, and simplifying, now yields (7.105), for $n \geq 2$.

By Theorem 5.33, the $n = 1$ case of (7.105) is established by showing that this case of the multiple sum in (7.105) is T_2, as defined by (5.110). This equality follows termwise by using (6.5) to compute the $i = 2$ case of (7.59). □

We next have the following theorem.

Theorem 7.20. *Let $\vartheta_2(0,q)$ and $\vartheta_3(0,q)$ be defined by* (1.2) *and* (1.1), *respectively. Let $n = 1, 2, 3, \ldots$. We then have*

$$\vartheta_2(0,q)^{2n^2}\vartheta_3(0,q)^{2n(n+1)} \cdot \left\{2n\left[1+24\sum_{r=1}^{\infty}\frac{rq^r}{1+q^r}\right]-\left[1-24\sum_{r=1}^{\infty}\frac{(2r-1)q^{2r-1}}{1+q^{2r-1}}\right]\right\}$$
$$= 4^n\frac{3}{n(2n+1)}\prod_{r=1}^{n}(2r-1)!^{-2}\sum_{\substack{y_1,\ldots,y_n\geq 1(\text{odd})\\ m_1>m_2>\cdots>m_n\geq 1(\text{odd})}}(-1)^{\frac{1}{2}(-n+y_1+\cdots+y_n)}$$
$$\cdot\left(m_1^2+\cdots+m_n^2\right)(m_1m_2\ldots m_n)^2 \cdot q^{\frac{1}{2}(m_1y_1+\cdots+m_ny_n)}\prod_{1\leq r<s\leq n}\left(m_r^2-m_s^2\right)^2. \tag{7.106}$$

Proof: We take $n \geq 2$ and apply Theorem 6.19 to Theorem 5.34. We utilize the same specializations as in (7.68)–(7.72), except that b_i is given by (7.92). It is immediate that (7.64) holds. In this application of the determinant expansion of (6.57) we also have (7.44), and (7.101)–(7.104).

Keeping in mind (7.44), (7.64), (7.69), (7.70), (7.72), (7.92), (7.101)–(7.104), noting that the inner sum in (6.57) consists of just one term corresponding to $T = I_n$, observing that (7.47) holds, relabelling $(2m_r - 1)$ and $(2y_r - 1)$ as m_r (odd) and y_r (odd), respectively, and simplifying, now yields (7.106), for $n \geq 2$.

By Theorem 5.34, the $n = 1$ case of (7.106) is established by showing that this case of the multiple sum in (7.106) is T_4, as defined by (5.110). This equality follows termwise by using (6.5) to compute the $i = 2$ case of (7.68). □

We next have the following theorem.

Theorem 7.21. *Let $\vartheta_2(0,q)$ and $\vartheta_3(0,q)$ be defined by* (1.2) *and* (1.1), *respectively. Let $n = 1, 2, 3, \ldots$. We then have*

$$\vartheta_2(0,q)^{2n(n-1)}\vartheta_3(0,q)^{2n^2} \cdot \left\{n\left[1+24\sum_{r=1}^{\infty}\frac{rq^r}{1+q^r}\right]-\left[1+24\sum_{r=1}^{\infty}\frac{rq^{2r}}{1+q^{2r}}\right]\right\}$$
$$= 4^{n^2}\frac{3}{2n(2n-1)}\prod_{r=1}^{n-1}(2r)!^{-2}\sum_{\substack{y_1,\ldots,y_n\geq 1(\text{odd})\\ m_1>m_2>\cdots>m_n\geq 1}}(-1)^{\frac{1}{2}(-n+y_1+\cdots+y_n)}$$
$$\cdot\left(m_1^2+\cdots+m_n^2\right)q^{m_1y_1+\cdots+m_ny_n}\prod_{1\leq r<s\leq n}\left(m_r^2-m_s^2\right)^2 \tag{7.107a}$$
$$+4^{n^2-1}\frac{3}{2n(2n-1)}\prod_{r=1}^{n-1}(2r)!^{-2}\sum_{\substack{y_1,\ldots,y_{n-1}\geq 1(\text{odd})\\ m_1>m_2>\cdots>m_{n-1}\geq 1}}(-1)^{\frac{1}{2}(-n+1+y_1+\cdots+y_{n-1})}$$

$$\cdot\,(m_1^2+\cdots+m_{n-1}^2)\,(m_1m_2\cdots m_{n-1})^4 \cdot q^{m_1y_1+\cdots+m_{n-1}y_{n-1}} \prod_{1\le r<s\le n-1}(m_r^2-m_s^2)^2, \tag{7.107b}$$

where the term in (7.107b) *is defined to be* 0 *if* $n=1$.

Proof: We apply Theorem 6.19 to Theorem 5.35. We first take $n\geq 2$ and apply Theorem 6.19 to the single $n\times n$ determinant in the right-hand-side of Eq. (5.276a) in Theorem 5.35. We utilize the same specializations as in (7.75)–(7.79), except that b_i is given by (7.92). It is immediate that (7.53) holds. In this application of the determinant expansion of (6.57) we also have (7.44), and (7.101)–(7.104).

Keeping in mind (7.44), (7.53), (7.76), (7.77), (7.79), (7.92), (7.101)–(7.104), noting that the inner sum in (6.57) consists of just one term corresponding to $T=I_n$, observing that $-m_r+2m_ry_r=m_r(2y_r-1)$, relabelling $(2y_r-1)$ as y_r (odd), and simplifying, now yields the term in (7.107a), for $n\geq 2$.

By Theorem 5.35, the $n=1$ case of (7.107a) is established by showing that this case of the multiple sum in (7.107a) is N_2, as defined by (5.128). This equality follows termwise by using (6.5) to compute the $i=2$ case of (7.75).

For $n\geq 3$, the derivation of the term in (7.107b) from (5.276b) is the same as for (7.107a), except that $n\mapsto n-1$, $c_i\mapsto 2i+2$, and $c_{\ell_1}+b_{j_1}\mapsto 4$.

By Theorem 5.35, the $n=2$ case of (7.107b) is established by showing that this case of the multiple sum in (7.107b) is N_6, as defined by (5.128). This equality follows termwise by using (6.5) to compute the $i=4$ case of (7.75).

If $n=1$, the term in (7.107b) is defined to be 0. □

We next have the following theorem.

Theorem 7.22. *Let* $\vartheta_2(0,q)$ *and* $\vartheta_3(0,q)$ *be defined by* (1.2) *and* (1.1), *respectively. Let* $n=1,2,3,\ldots$. *We then have*

$$\vartheta_2(0,q)^{2n(n+1)}\vartheta_3(0,q)^{2n^2}\cdot\left\{n\left[1+24\sum_{r=1}^{\infty}\frac{rq^r}{1+q^r}\right]+\left[1+24\sum_{r=1}^{\infty}\frac{rq^{2r}}{1+q^{2r}}\right]\right\}$$

$$=4^{n(n+1)}\frac{6}{n(2n+1)}\prod_{r=1}^{n}(2r-1)!^{-2}\sum_{\substack{y_1,\ldots,y_n\geq 1(\text{odd})\\ m_1>m_2>\cdots>m_n\geq 1}}(-1)^{\frac{1}{2}(-n+y_1+\cdots+y_n)}$$

$$\cdot\,(m_1^2+\cdots+m_n^2)\,(m_1m_2\ldots m_n)^2 \cdot q^{m_1y_1+\cdots+m_ny_n}\prod_{1\le r<s\le n}(m_r^2-m_s^2)^2. \tag{7.108}$$

Proof: We take $n\geq 2$ and apply Theorem 6.19 to Theorem 5.36. We utilize the same specializations as in (7.81)–(7.85), except that b_i is given by (7.92). It is immediate that (7.53) holds. In this application of the determinant expansion of (6.57) we also have (7.44), and (7.101)–(7.104).

Keeping in mind (7.44), (7.53), (7.82), (7.83), (7.85), (7.92), (7.101)–(7.104), noting that the inner sum in (6.57) consists of just one term corresponding to $T = I_n$, observing that $-m_r + 2m_r y_r = m_r(2y_r - 1)$, relabelling $(2y_r - 1)$ as y_r (odd), and simplifying, now yields (7.108), for $n \geq 2$.

By Theorem 5.36, the $n = 1$ case of (7.108) is established by showing that this case of the multiple sum in (7.108) is N_4, as defined by (5.128). This equality follows termwise by using (6.5) to compute the $i = 2$ case of (7.81). □

Remark. Note that the terms in the multiple sums in (7.99), (7.100), (7.105), (7.106), (7.107a), and (7.108) are simply $(m_1^2 + \cdots + m_n^2)$ times the corresponding terms in (7.36), (7.37), (7.58), (7.67), (7.74a), and (7.80), respectively.

Chan's paper [41], which utilized Dedekind's η-function transformation formula and Hecke's correspondence between Fourier series and Dirichlet series to prove the equivalence of some partition identities of Ramanujan, motivates a direct relationship between Theorems 7.1 and 7.2 and the simpler identities in Theorem 7.5. This relationship is a consequence of

$$\vartheta_3(0, -q) = (-i\tau)^{-1/2} \vartheta_2(0, q_1), \tag{7.109}$$

where

$$q := e^{i\pi\tau} \quad \text{and} \quad q_1 := e^{-i\pi/\tau}. \tag{7.110}$$

To see (7.109), take $z = 0$ in Jacobi's transformation

$$\vartheta_4(\tau z \,|\, \tau) = (-i\tau)^{-1/2} e^{-i\tau z^2/\pi} \vartheta_2\left(z \middle| \tfrac{-1}{\tau}\right) \tag{7.111}$$

in [135, Eq. (1.7.15), p. 17] of the theta functions ϑ_4 and ϑ_2. Equation (7.109) implies that

$$\vartheta_3(0, -q)^{4n^2} = (-1)^n \tau^{-2n^2} \vartheta_2(0, q_1)^{4n^2}, \tag{7.112}$$

and

$$\vartheta_3(0, -q)^{4n(n+1)} = \tau^{-2n(n+1)} \vartheta_2(0, q_1)^{4n(n+1)}, \tag{7.113}$$

where q and q_1 are given by (7.110).

Applying Theorem 7.5 to the right-hand sides of (7.112) and (7.113) now leads to the following corollary.

Corollary 7.23. *Let* $\vartheta_3(0, q)$ *be defined by* (1.1). *Let* q *and* q_1 *be given by* (7.110). *Let* $n = 1, 2, 3, \ldots$. *We then have*

$$\vartheta_3(0, -q)^{4n^2} = (-1)^n 4^{n(n+1)} \tau^{-2n^2} \prod_{r=1}^{2n-1} (r!)^{-1} \sum_{\substack{y_1, \ldots, y_n \geq 1 (\text{odd}) \\ m_1 > m_2 > \cdots > m_n \geq 1 (\text{odd})}} m_1 m_2 \ldots m_n \cdot q_1^{m_1 y_1 + \cdots + m_n y_n} \prod_{1 \leq r < s \leq n} \left(m_r^2 - m_s^2\right)^2, \tag{7.114}$$

and

$$\vartheta_3(0,-q)^{4n(n+1)} = 2^{n(4n+5)}\tau^{-2n(n+1)} \prod_{r=1}^{2n}(r!)^{-1} \sum_{\substack{y_1,\ldots,y_n \geq 1(\text{odd}) \\ m_1>m_2>\cdots>m_n\geq 1}} (m_1 m_2 \ldots m_n)^3 \cdot q_1^{2(m_1y_1+\cdots+m_ny_n)} \prod_{1\leq r<s\leq n} (m_r^2 - m_s^2)^2. \qquad (7.115)$$

To obtain an identity just involving multiple sums, replace the products $\vartheta_3(0,-q)^{4n^2}$ and $\vartheta_3(0,-q)^{4n(n+1)}$ in (7.114) and (7.115) by the sums in (7.1) and (7.11).

A Lambert series version of Corollary 7.23 follows by applying Theorems 5.4, 5.6, and 5.11 to the products in (7.112) and (7.113). The analysis in [41, 42] suggests that an extension of the Hecke correspondence between Fourier series and Dirichlet series, applied to this Lambert series version of Corollary 7.23, may lead to functional equations and evaluations for the corresponding multiple L-series.

The classical techniques in [90, pp. 96–101] and [28, pp. 288–305] applied to Theorems 7.1 through 7.22 lead to corresponding infinite families of lattice sum transformations. We illustrate this procedure by applying the Mellin transform to Theorems 7.1, 7.2, 7.5, and the two special cases in Corollaries 8.1 and 8.2. These computations involve the specialized theta functions $\vartheta_2(q) := \vartheta_2(0,q)$, $\vartheta_3(q) := \vartheta_3(0,q)$, $\vartheta_4(q) := \vartheta_3(0,-q)$, and the Mellin transform $Mf(s)$ given by

$$Mf(s) \equiv M_s(f) := \int_0^\infty f(t)t^{s-1}\,dt. \qquad (7.116)$$

We make frequent use of

$$M_s(e^{-nt}) = n^{-s}\cdot\Gamma(s), \qquad (7.117)$$

where $\Gamma(s)$ is the gamma function.

We first study the m-dimensional cubic lattice sums $b_m(2s)$ defined by

$$b_m(2s) := \sum_{\substack{(i_1,\ldots,i_m)\neq(0,\ldots,0) \\ -\infty<i_1,\ldots,i_m<\infty}} \frac{(-1)^{i_1+\cdots+i_m}}{(i_1^2+\cdots+i_m^2)^s}, \qquad (7.118)$$

where $\text{Re}(s) > 0$. The convergence of these general sums is discussed in [28, Example 1 and 2, pp. 290–291]. Sums of the form $b_3(2s)$ occur naturally in chemistry. For example, $b_3(1)$ may be viewed as the potential or Coulomb sum at the origin of a cubic lattice with alternating unit charges at all nonzero lattice points. This is essentially an idealization of a rocksalt crystal. The quantity $b_3(1)$ is known as *Madelung's constant* for NaCl. Different crystals lead to different lattice sums.

Applying the Mellin transform termwise to both sides of the $q = e^{-t}$ case of (7.1) yields the following theorem.

Theorem 7.24. *Let $n = 1, 2, 3, \ldots$. We then have the formal identity*

$$b_{4n^2}(2s) = \sum_{p=1}^{n} (-1)^p 2^{2n^2+n} \prod_{r=1}^{2n-1} (r!)^{-1} \sum_{\substack{y_1,\ldots,y_p \geq 1 \\ m_1 > m_2 > \cdots > m_p \geq 1}} (-1)^{y_1+\cdots+y_p}$$
$$\cdot (-1)^{m_1+\cdots+m_p} (m_1 y_1 + \cdots + m_p y_p)^{-s} \prod_{1 \leq r < s \leq p} (m_r^2 - m_s^2)^2$$
$$\cdot (m_1 m_2 \ldots m_p) \sum_{\substack{\emptyset \subset S, T \subseteq I_n \\ \|S\| = \|T\| = p}} (-1)^{\Sigma(S)+\Sigma(T)} \cdot \det\left(D_{n-p,S^c,T^c}\right)$$
$$\cdot (m_1 m_2 \ldots m_p)^{2\ell_1 + 2j_1 - 4} s_\lambda\left(m_1^2, \ldots, m_p^2\right) s_\nu\left(m_1^2, \ldots, m_p^2\right), \quad (7.119)$$

where the same assumptions hold as in Theorem 7.1, and $b_{4n^2}(2s)$ is determined by (7.118).

Similarly, applying the Mellin transform termwise to both sides of the $q = e^{-t}$ case of (7.11) yields the following theorem.

Theorem 7.25. *Let $n = 1, 2, 3, \ldots$. We then have the formal identity*

$$b_{4n(n+1)}(2s) = \sum_{p=1}^{n} (-1)^{n-p} 2^{2n^2+3n} \prod_{r=1}^{2n} (r!)^{-1} \sum_{\substack{y_1,\ldots,y_p \geq 1 \\ m_1 > m_2 > \cdots > m_p \geq 1}} (-1)^{m_1+\cdots+m_p}$$
$$\cdot (m_1 y_1 + \cdots + m_p y_p)^{-s} (m_1 m_2 \ldots m_p)^3 \prod_{1 \leq r < s \leq p} (m_r^2 - m_s^2)^2$$
$$\cdot \sum_{\substack{\emptyset \subset S, T \subseteq I_n \\ \|S\| = \|T\| = p}} (-1)^{\Sigma(S)+\Sigma(T)} \cdot \det\left(D_{n-p,S^c,T^c}\right)$$
$$\cdot (m_1 m_2 \ldots m_p)^{2\ell_1 + 2j_1 - 4} s_\lambda\left(m_1^2, \ldots, m_p^2\right) s_\nu\left(m_1^2, \ldots, m_p^2\right), \quad (7.120)$$

where the same assumptions hold as in Theorem 7.2, and $b_{4n(n+1)}(2s)$ is determined by (7.118).

It appears that the convergence conditions $\text{Re}(s) > 4n - 3$ and $\text{Re}(s) > 4n - 1$ are sufficient for Theorems 7.24 and 7.25, respectively.

We next study the m-dimensional cubic lattice sums $c_m(2s)$ defined by

$$c_m(2s) := \sum_{-\infty < i_1, \ldots, i_m < \infty} [(i_1 + 1/2)^2 + \cdots + (i_m + 1/2)^2]^{-s}, \quad (7.121)$$

where $\text{Re}(s) > m/2$. This convergence condition is determined by utilizing the estimates for $r_m(N)$ in [94, Eq. (9.20), p. 122] to compare (7.121) with $\zeta(s - \frac{m}{2} + 1)$.

Applying the Mellin transform termwise to both sides of the $q = e^{-t}$ cases of (7.36) and (7.37) yields the following theorem.

Theorem 7.26. *Let $n = 1, 2, 3, \ldots$, and let $c_{4n^2}(2s)$ and $c_{4n(n+1)}(2s)$ be determined by (7.121). We then have the formal identities*

$$c_{4n^2}(2s) = 4^{n(n+1)} \cdot \prod_{r=1}^{2n-1} (r!)^{-1} \sum_{\substack{y_1,\ldots,y_n \geq 1 (\text{odd}) \\ m_1 > m_2 > \cdots > m_n \geq 1 (\text{odd})}} m_1 m_2 \ldots m_n \\ \cdot (m_1 y_1 + \cdots + m_n y_n)^{-s} \prod_{1 \leq r < s \leq n} (m_r^2 - m_s^2)^2, \tag{7.122}$$

and

$$c_{4n(n+1)}(2s) = 2^{-s} 2^{4n^2+5n} \cdot \prod_{r=1}^{2n} (r!)^{-1} \sum_{\substack{y_1,\ldots,y_n \geq 1 (\text{odd}) \\ m_1 > m_2 > \cdots > m_n \geq 1}} (m_1 m_2 \ldots m_n)^3 \\ \cdot (m_1 y_1 + \cdots + m_n y_n)^{-s} \prod_{1 \leq r < s \leq n} (m_r^2 - m_s^2)^2. \tag{7.123}$$

It appears that the convergence conditions $\text{Re}(s) > 2n^2$ and $\text{Re}(s) > 2n(n+1)$ are sufficient for (7.122) and (7.123), respectively. Furthermore, the right-hand sides of (7.122) and (7.123) may be an analytic continuation of the left-hand sides to $\text{Re}(s) > 4n - 2$ and $\text{Re}(s) > 4n$, respectively.

We now make use of Corollaries 8.1 and 8.2 to write down simplified versions of the $n = 2$ cases of Theorems 7.24 and 7.25.

We have the following corollaries.

Corollary 7.27. *Let $b_{16}(2s)$ be determined by (7.118). We then have the formal identity*

$$b_{16}(2s) = -\tfrac{2^5}{3} \sum_{y_1 \geq 1} (-1)^{y_1} y_1^{-s} \cdot \sum_{m_1 \geq 1} (-1)^{m_1} (1 + m_1^2 + m_1^4) m_1^{1-s} \\ + \tfrac{2^8}{3} \sum_{\substack{y_1, y_2 \geq 1 \\ m_1 > m_2 \geq 1}} (-1)^{y_1 + y_2 + m_1 + m_2} (m_1 m_2)(m_1^2 - m_2^2)^2 (m_1 y_1 + m_2 y_2)^{-s}. \tag{7.124}$$

Corollary 7.28. *Let $b_{24}(2s)$ be determined by (7.118). We then have the formal identity*

$$b_{24}(2s) = +\tfrac{2^4}{3^2} \cdot \zeta(s) \cdot \sum_{m_1 \geq 1} (-1)^{m_1} (17 + 8m_1^2 + 2m_1^4) m_1^{3-s} \\ + \tfrac{2^9}{3^2} \sum_{\substack{y_1, y_2 \geq 1 \\ m_1 > m_2 \geq 1}} (-1)^{m_1 + m_2} (m_1 m_2)^3 (m_1^2 - m_2^2)^2 (m_1 y_1 + m_2 y_2)^{-s}. \tag{7.125}$$

It appears that the convergence conditions $\mathrm{Re}(s) > 5$ and $\mathrm{Re}(s) > 7$ are sufficient for Corollaries 7.27 and 7.28, respectively.

The analysis involving Theorems 7.24–7.26 may have applications to the work in [32].

8. The 36 and 48 squares identities

In this section we write down the $n = 2$ and $n = 3$ cases of Theorems 7.1 and 7.2. These results provide explicit multiple power series formulas for 16, 24, 36, and 48 squares.

The first two corollaries give the $n = 2$ cases of Theorems 7.1 and 7.2, respectively.

Corollary 8.1. *Let* $\vartheta_3(0, -q)$ *be determined by* (1.1). *Then*

$$\vartheta_3(0,-q)^{16} = 1 - \frac{2^5}{3} \sum_{y_1, m_1 \geq 1} (-1)^{y_1+m_1} m_1 \left(1 + m_1^2 + m_1^4\right) q^{m_1 y_1} \tag{8.1a}$$

$$+ \frac{2^8}{3} \sum_{\substack{y_1, y_2 \geq 1 \\ m_1 > m_2 \geq 1}} (-1)^{y_1+y_2+m_1+m_2} (m_1 m_2) \left(m_1^2 - m_2^2\right)^2 q^{m_1 y_1 + m_2 y_2}. \tag{8.1b}$$

Corollary 8.2. *Let* $\vartheta_3(0, -q)$ *be determined by* (1.1). *Then*

$$\vartheta_3(0,-q)^{24} = 1 + \frac{2^4}{3^2} \sum_{y_1, m_1 \geq 1} (-1)^{m_1} m_1^3 \left(17 + 8m_1^2 + 2m_1^4\right) q^{m_1 y_1} \tag{8.2a}$$

$$+ \frac{2^9}{3^2} \sum_{\substack{y_1, y_2 \geq 1 \\ m_1 > m_2 \geq 1}} (-1)^{m_1+m_2} (m_1 m_2)^3 \left(m_1^2 - m_2^2\right)^2 q^{m_1 y_1 + m_2 y_2}. \tag{8.2b}$$

The next two corollaries give the $n = 3$ cases of Theorems 7.1 and 7.2, respectively.

Corollary 8.3. *Let* $\vartheta_3(0, -q)$ *be determined by* (1.1). *Then*

$$\begin{aligned}
&\vartheta_3(0,-q)^{36} \\
&\quad = 1 - \frac{2^3}{3^2 \cdot 5} \sum_{y_1, m_1 \geq 1} (-1)^{y_1+m_1} q^{m_1 y_1} m_1 \\
&\qquad \cdot \left[69 + 120m_1^2 + 172m_1^4 + 40m_1^6 + 4m_1^8\right] \\
&\quad + \frac{2^9}{3^3 \cdot 5} \sum_{\substack{y_1, y_2 \geq 1 \\ m_1 > m_2 \geq 1}} (-1)^{y_1+y_2+m_1+m_2} q^{m_1 y_1 + m_2 y_2} (m_1 m_2) \left(m_1^2 - m_2^2\right)^2 \\
&\qquad \cdot \left[62 + 17m_1^2 + 17m_2^2 + 2m_1^4 + 2m_2^4 + 8m_1^2 m_2^2 + 2m_1^4 m_2^2 + 2m_1^2 m_2^4 + 2m_1^4 m_2^4\right] \\
&\quad - \frac{2^{13}}{3^3 \cdot 5} \sum_{\substack{y_1, y_2, y_3 \geq 1 \\ m_1 > m_2 > m_3 \geq 1}} (-1)^{(y_1+y_2+y_3)+(m_1+m_2+m_3)} q^{m_1 y_1 + m_2 y_2 + m_3 y_3} \\
&\qquad \cdot (m_1 m_2 m_3) \prod_{1 \leq r < s \leq 3} \left(m_r^2 - m_s^2\right)^2.
\end{aligned} \tag{8.3}$$

Corollary 8.4. *Let $\vartheta_3(0,-q)$ be determined by* (1.1). *Then*

$$\begin{aligned}
\vartheta_3(0,-q)^{48} \\
= 1 &+ \tfrac{2^5}{3^3\cdot5^2} \sum_{y_1,m_1\geq1} (-1)^{m_1} q^{m_1y_1} m_1^3 \\
&\cdot \left[902 + 760m_1^2 + 321m_1^4 + 40m_1^6 + 2m_1^8\right] \\
&+ \tfrac{2^{10}}{3^5\cdot5^2} \sum_{\substack{y_1,y_2\geq1 \\ m_1>m_2\geq1}} (-1)^{m_1+m_2} q^{m_1y_1+m_2y_2} (m_1m_2)^3 \left(m_1^2-m_2^2\right)^2 \\
&\cdot \left[1382+248m_1^2+248m_2^2+17m_1^4+17m_2^4+68m_1^2m_2^2+8m_1^4m_2^2+8m_1^2m_2^4+2m_1^4m_2^4\right] \\
&+ \tfrac{2^{15}}{3^5\cdot5^2} \sum_{\substack{y_1,y_2,y_3\geq1 \\ m_1>m_2>m_3\geq1}} (-1)^{m_1+m_2+m_3} q^{m_1y_1+m_2y_2+m_3y_3} \cdot (m_1m_2m_3)^3 \prod_{1\leq r<s\leq3} \left(m_r^2-m_s^2\right)^2.
\end{aligned} \tag{8.4}$$

Acknowledgment

The author would like to thank George Andrews, Richard Askey, Bruce Berndt, Mourad Ismail, and Christian Krattenthaler for some of the references to the literature. In addition, he thanks Krishnaswami Alladi and Kluwer Academic Publishers for also putting this paper out in hard cover form in the book series Developments in Mathematics, Vol. 5.

Note

1. In Jones and Thron [119, pp. 244–246] Theorem 7.14, χ_1 is incorrectly given as c_1.

References

1. N.H. Abel, "Recherches sur les fonctions elliptiques," *J. Reine Angew. Math.* **2** (1827), 101–181; reprinted in Œuvres Complètes T1, Grondahl and Son, Christiania, 1881, pp. 263–388; reprinted by Johnson Reprint Corporation, New York, 1965.
2. W.A. Al-Salam and L. Carlitz, "Some determinants of Bernoulli, Euler, and related numbers," *Portugal. Math.* **18**(2) (1959), 91–99.
3. K. Ananda-Rau, "On the representation of a number as the sum of an even number of squares," *J. Madras Univ. Sect. B* **24** (1954), 61–89.
4. G.E. Andrews, "Applications of basic hypergeometric functions," *SIAM Rev.* **16** (1974), 441–484.
5. G.E. Andrews, "*q*-Series: Their development and application in analysis, number theory, combinatorics, physics and computer algebra," in *NSF CBMS Regional Conference Series*, Vol. 66, 1986.
6. G.E. Andrews, R. Askey, and R. Roy, "Special functions," in *Encyclopedia of Mathematics and its Applications*, Vol. 71 (G.-C. Rota, ed.), Cambridge University Press, Cambridge, 1999.
7. G.E. Andrews and B.C. Berndt, *Ramanujan's Lost Notebook*, Part I, Springer-Verlag, New York, in preparation.
8. T.M. Apostol, *Modular Functions and Dirichlet Series in Number Theory*, Vol. 41 of Graduate Texts in Mathematics, Springer-Verlag, New York, 1976.

9. R. Askey and M.E.H. Ismail, "Recurrence relations, continued fractions and orthogonal polynomials," *Mem. Amer. Math. Soc.* **300** (1984), 108 pp.
10. H. Au-Yang and J.H.H. Perk, "Critical correlations in a Z-invariant inhomogeneous Ising model," *Phys. A* **144** (1987), 44–104.
11. I.G. Bashmakova, "Diophantus and diophantine equations," Vol. 20 of The Dolciani Mathematical Expositions, Mathematical Association of America, Washington, DC, 1997, xvi+90 pp.; translated from the 1972 Russian original by Abe Shenitzer and updated by Joseph Silverman.
12. I.G. Bashmakova and G.S. Smirnova, "The birth of literal algebra," *Amer. Math. Monthly* **106** (1999), 57–66; translated from the Russian and edited by Abe Shenitzer.
13. E.F. Beckenbach, W. Seidel, and O. Szász, "Recurrent determinants of Legendre and of ultraspherical polynomials," *Duke Math. J.* **18** (1951), 1–10.
14. E.T. Bell, "On the number of representations of $2n$ as a sum of $2r$ squares," *Bull. Amer. Math. Soc.* **26** (1919), 19–25.
15. E.T. Bell, "Theta expansions useful in arithmetic," *The Messenger of Mathematics (New Series)* **53** (1924), 166–176.
16. E.T. Bell, "On the power series for elliptic functions," *Trans. Amer. Math. Soc.* **36** (1934), 841–852.
17. E.T. Bell, "The arithmetical function $M(n, f, g)$ and its associates connected with elliptic power series," *Amer. J. Math.* **58** (1936), 759–768.
18. E.T. Bell, "Polynomial approximations for elliptic functions," *Trans. Amer. Math. Soc.* **44** (1938), 47–57.
19. C. Berg and G. Valent, "The Nevanlinna parameterization for some indeterminate Stieltjes moment problems associated with birth and death processes," *Methods Appl. Anal.* **1** (1994), 169–209.
20. B.C. Berndt, *Ramanujan's Notebooks*, Part II, Springer-Verlag, New York, 1989.
21. B.C. Berndt, *Ramanujan's Notebooks*, Part III, Springer-Verlag, New York, 1991.
22. B.C. Berndt, "Ramanujan's theory of theta-functions," in *Theta Functions From the Classical to the Modern* (M. Ram Murty, ed.), Vol. 1 of CRM Proceedings & Lecture Notes, American Mathematical Society, Providence, RI, 1993, 1–63.
23. B.C. Berndt, *Ramanujan's Notebooks*, Part V, Springer-Verlag, New York, 1998.
24. B.C. Berndt, "Fragments by Ramanujan on Lambert Series," in *Number Theory and Its Applications* (Kyoto, 1997) (K. Györy and S. Kanemitsu, eds.), Vol. 2 of Dev. Math., Kluwer Academic Publishers, Dordrecht, 1999, pp. 35–49.
25. M. Bhaskaran, "A plausible reconstruction of Ramanujan's proof of his formula for $\vartheta^{4s}(q)$," in *Ananda Rau Memorial Volume*, Publications of the Ramanujan Institute, No. 1., Ramanujan Institute, Madras, 1969, pp. 25–33.
26. M.N. Bleicher and M.I. Knopp, "Lattice points in a sphere," *Acta Arith.* **10** (1964/1965), 369–376.
27. F. van der Blij[i], "The function $\tau(n)$ of S. Ramanujan," *Math. Student* **18** (1950), 83–99.
28. J.M. Borwein and P.B. Borwein, *Pi and the AGM*, John Wiley & Sons, New York, 1987.
29. D.M. Bressoud, "Proofs and confirmations. The story of the alternating sign matrix conjecture," in *MAA Spectrum*, Mathematical Association of America, Washington, DC/Cambridge University Press, Cambridge, 1999, pp. 245–256.
30. D.M. Bressoud and J. Propp, "How the alternating sign matrix conjecture was solved," *Notices Amer. Math. Soc.* **46** (1999), 637–646.
31. C. Brezinski, *History of Continued Fractions and Padé Approximants*, Vol. 12 of Springer Series in Computational Mathematics, Springer-Verlag, New York, 1991.
32. D.J. Broadhurst, "On the enumeration of irreducible k-fold Euler sums and their roles in knot theory and field theory," *J. Math. Phys.*, to appear.
33. V. Bulygin, "Sur une application des fonctions elliptiques au problème de représentation des nombres entiers par une somme de carrés," *Bull. Acad. Imp. Sci. St. Petersbourg Ser. VI* **8** (1914), 389–404; B. Boulyguine, "Sur la représentation d'un nombre entier par une somme de carrés," *Comptes Rendus Paris* **158** (1914), 328–330.
34. V. Bulygin (B. Boulyguine), "Sur la représentation d'un nombre entier par une somme de carrés," *Comptes Rendus Paris* **161** (1915), 28–30.
35. J.L. Burchnall, "An algebraic property of the classical polynomials," *Proc. London Math. Soc.* **1**(3) (1951), 232–240.

36. L. Carlitz, "Hankel determinants and Bernoulli numbers," *Tôhoku Math. J.* **5**(2) (1954), 272–276.
37. L. Carlitz, "Note on sums of 4 and 6 squares," *Proc. Amer. Math. Soc.* **8** (1957), 120–124.
38. L. Carlitz, "Some orthogonal polynomials related to elliptic functions," *Duke Math. J.* **27** (1960), 443–459.
39. L. Carlitz, "Bulygin's method for sums of squares. The arithmetical theory of quadratic forms, I," in *Proc. Conf.*, Louisiana State Univ., Baton Rouge, LA, 1972 (dedicated to Louis Joel Mordell); also in *J. Number Theory* **5** (1973), 405–412.
40. R. Chalkley, "A persymmetric determinant," *J. Math. Anal. Appl.* **187** (1994), 107–117.
41. H.H. Chan, "On the equivalence of Ramanujan's partition identities and a connection with the Rogers-Ramanujan continued fraction," *J. Math. Anal. Appl.* **198** (1996), 111–120.
42. H.H. Chan, private communication, August 1996.
43. K. Chandrasekharan, *Elliptic Functions*, Vol. 281 of Grundlehren Math. Wiss, Springer-Verlag, Berlin, 1985.
44. T.S. Chihara, *An Introduction to Orthogonal Polynomials*, Vol. 13 of Mathematics and Its Applications Gordon and Breach, New York, 1978.
45. S.H. Choi and D. Gouyou-Beauchamps, "Enumération de tableaux de Young semi-standard," in *Series Formelles et Combinatoire Algebrique: Actés du Colloque* (M. Delest, G. Jacob, and P. Leroux, eds.), Université Bordeaux I, 2–4 May, 1991, pp. 229–243.
46. D.V. Chudnovsky and G.V. Chudnovsky, "Computational problems in arithmetic of linear differential equations. Some diophantine applications," *Number Theory New York*, 1985–88 (D. & G. Chudnovsky, H. Cohn, and M. Nathanson, eds.), Vol. 1383 of Lecture Notes in Math., Springer-Verlag, New York, 1989, pp. 12–49.
47. D.V. Chudnovsky and G.V. Chudnovsky, "Hypergeometric and modular function identities, and new rational approximations to and continued fraction expansions of classical constants and functions," *A Tribute to Emil Grosswald: Number theory and related analysis* (M. Knopp and M. Sheingorn, eds.), Vol. 143 of Contemporary Mathematics, American Mathematical Society, Providence, RI, 1993, pp. 117–162.
48. L. Comtet, *Advanced Combinatorics*, D. Reidel Pub. Co., Dordrecht-Holland/Boston-USA, 1974.
49. E. Conrad, "A note on certain continued fraction expansions of Laplace transforms of Dumont's bimodular Jacobi elliptic functions," preprint.
50. E. Conrad, *A Handbook of Jacobi Elliptic Functions*, Class notes (1996), preprint.
51. J.H. Conway and N.J.A. Sloane, *Sphere Packings, Lattices and Groups*, 3rd edn. (with additional contributions by E. Bannai, R.E. Borcherds, J. Leech, S.P. Norton, A.M. Odlyzko, R.A. Parker, L. Queen, and B.B. Venkov), Vol. 290 of Grundlehren der Mathematischen Wissenschaften, Springer-Verlag, New York, 1999.
52. T.L. Curtright and C.B. Thorn, "Symmetry patterns in the mass spectra of dual string models," *Nuclear Phys. B* **274** (1986), 520–558.
53. T.W. Cusick, "Identities involving powers of persymmetric determinants," *Proc. Cambridge Philos. Soc.* **65** (1969), 371–376.
54. H. Datta, "On the theory of continued fractions," *Proc. Edinburgh Math. Soc.* **34** (1916), 109–132.
55. P. Delsarte, "Nombres de Bell et polynômes de Charlier," *C.R. Acad. Sc. Paris* (Series A) **287** (1978), 271–273.
56. L.E. Dickson, *History of the Theory of Numbers*, Vol. 2, Chelsea, New York, 1966.
57. A.C. Dixon, "On the doubly periodic functions arising out of the curve $x^3 + y^3 - 3\alpha xy = 1$," *The Quarterly Journal of Pure and Applied Mathematics* **24** (1890), 167–233.
58. D. Dumont, "Une approche combinatoire des fonctions elliptiques de Jacobi," *Adv. in Math.* **41** (1981), 1–39.
59. D. Dumont, "Pics de cycle et dérivées partielles," *Séminaire Lotharingien de Combinatoire* **13**(B13a) (1985), 19 pp.
60. D. Dumont, "Le paramétrage de la courbe d'équation $x^3 + y^3 = 1$" (Une introduction élémentaire aux fonctions elliptiques), preprint (May 1988).
61. F.J. Dyson, "Missed opportunities," *Bull. Amer. Math. Soc.* **78** (1972), 635–653.
62. R. Ehrenborg, "The Hankel determinant of exponential polynomials," *Amer. Math. Monthly* **107** (2000), 557–560.
63. A. Erdélyi (with A. Magnus, F. Oberhettinger, and F. Tricomi), *Higher Transcendental Functions*, Bateman Manuscript Project (A. Erdélyi, ed.), Vol. II, McGraw-Hill Book Co., New York, 1953; reissued by Robert E. Krieger Pub. Co., Malabar, Florida, 1981 and 1985.

64. T. Estermann, "On the representations of a number as a sum of squares," *Acta Arith.* **2** (1936), 47–79.
65. L. Euler, "De fractionibus continuis dissertatio," *Comm. Acad. Sci. Imp. St. Pétersbourg* **9** (1737), 98–137; reprinted in Works. 1911–. Leonhardi Euleri Opera Omnia (F. Rudio, A. Krazer, and P. Stackel, eds.), Ser. I, Vol. 14 (C. Boehm and G. Faber, eds.), B.G. Teubner, Lipsiae 1925, pp. 187–215; see also, "An essay on continued fractions," *Math. Systems Theory* **18** (1985), 295–328; translated from the Latin by Myra F. Wyman and Bostwick F. Wyman.
66. L. Euler, "De fractionibus continuis observationes," *Comm. Acad. Sci. Imp. St. Pétersbourg* **11** (1739), 32–81; reprinted in Works. 1911–. Leonhardi Euleri Opera Omnia (F. Rudio, A. Krazer, and P. Stackel, eds.), Ser. I, Vol. 14 (C. Boehm and G. Faber, eds.), B.G. Teubner, Lipsiae, 1925, pp. 291–349.
67. L. Euler, *Introductio in Analysin Infinitorum*, Vol. I, Marcum-Michaelem Bousquet, Lausanne, 1748; reprinted in Works. 1911–. Leonhardi Euleri Opera Omnia (F. Rudio, A. Krazer, and P. Stackel, eds.), Ser. I, Vol. 8 (A. Krazer and F. Rudio, eds.), B.G. Teubner, Lipsiae, 1922, pp. 1–392, (see bibliographie on page b*); see also, *Introduction to Analysis of the Infinite: Book I*, Springer-Verlag, New York, 1988; translated from the Latin by John D. Blanton.
68. L. Euler, *De transformatione serierum in fractiones continuas: ubi simul haec theoria non mediocriter amplificatur*, Opuscula Analytica, t. ii, Petropoli: Typis Academiae Imperialis Scientiarum (1783–1785), 1785, pp. 138–177; reprinted in Works. 1911–. Leonhardi Euleri Opera Omnia (F. Rudio, A. Krazer, A. Speiser, and L.G. du Pasquier. eds.), Ser. I, Vol. 15 (G. Faber, ed.), B.G. Teubner, Lipsiae, 1927, pp. 661–700.
69. P. Flajolet, "Combinatorial aspects of continued fractions," *Discrete Math.* **32** (1980), 125–161.
70. P. Flajolet, "On congruences and continued fractions for some classical combinatorial quantities," *Discrete Math.* **41** (1982), 145–153.
71. P. Flajolet and J. Françon, "Elliptic functions, continued fractions and doubled permutations," *European J. Combin.* **10** (1989), 235–241.
72. F.G. Frobenius, "Über Relationen zwischen den Näherungsbrüchen von Potenzreihen," *J. Reine Angew. Math.* **90** (1881), 1–17; reprinted in *Frobenius' Gesammelte Abhandlungen* (J.-P. Serre, ed.), Vol. 2, Springer-Verlag, Berlin, 1968, pp. 47–63.
73. F.G. Frobenius and L. Stickelberger, "Zur Theorie der elliptischen Functionen," *J. Reine Angew. Math.* **83** (1877), 175–179; reprinted in *Frobenius' Gesammelte Abhandlungen* (J.-P. Serre, ed.), Vol. 1, Springer-Verlag, Berlin, 1968, pp. 335–339.
74. F.G. Frobenius and L. Stickelberger, "Über die Addition und Multiplication der elliptischen Functionen," *J. Reine Angew. Math.* **88** (1880), 146–184; reprinted in *Frobenius' Gesammelte Abhandlungen* (J.-P. Serre, ed.), Vol. 1, Springer-Verlag, Berlin, 1968, pp. 612–650.
75. M. Fulmek and C. Krattenthaler, "The number of rhombus tilings of a symmetric hexagon which contain a fixed rhombus on the symmetry axis, II," *European J. Combin.* **21** (2000), 601–640.
76. H. Garland, "Dedekind's η-function and the cohomology of infinite dimensional Lie algebras," *Proc. Nat. Acad. Sci., U.S.A.* **72** (1975), 2493–2495.
77. H. Garland and J. Lepowsky, "Lie algebra homology and the Macdonald-Kac formulas," *Invent. Math.* **34** (1976), 37–76.
78. F. Garvan, private communication, March 1997.
79. G. Gasper and M. Rahman, "Basic hypergeometric series," in *Encyclopedia of Mathematics and its Applications*, Vol. 35 (G.-C. Rota, ed.), Cambridge University Press, Cambridge, 1990.
80. J. Geronimus, "On some persymmetric determinants," *Proc. Roy. Soc. Edinburgh* **50** (1930), 304–309.
81. J. Geronimus, "On some persymmetric determinants formed by the polynomials of M. Appell," *J. London Math. Soc.* **6** (1931), 55–59.
82. I. Gessel and G. Viennot, "Binomial determinants, paths, and hook length formulae," *Adv. in Math.* **58** (1985), 300–321.
83. F. Gesztesy and R. Weikard, "Elliptic algebro-geometric solutions of the KdV and AKNS hierarchies—An analytic approach," *Bull. Amer. Math. Soc.* (N.S.) **35** (1998), 271–317.
84. J.W.L. Glaisher, "On the square of the series in which the coefficients are the sums of the divisors of the exponents," *Mess. Math., New Series* **14** (1884-85), 156–163; reprinted in J.W.L. Glaisher, *Mathematical Papers*, Chiefly Connected with the q-series in Elliptic Functions (1883–1885), Cambridge, W. Metcalfe and Son, Trinity Street, 1885, pp. 371–379.

85. J.W.L. Glaisher, "On the numbers of representations of a number as a sum of $2r$ squares, where $2r$ does not exceed eighteen," *Proc. London Math. Soc.* **5**(2) (1907), 479–490.
86. J.W.L. Glaisher, "On the representations of a number as the sum of two, four, six, eight, ten, and twelve squares," *Quart. J. Pure and Appl. Math. Oxford* **38** (1907), 1–62.
87. J.W.L. Glaisher, "On the representations of a number as the sum of fourteen and sixteen squares," *Quart. J. Pure and Appl. Math. Oxford* **38** (1907), 178–236.
88. J.W.L. Glaisher, "On the representations of a number as the sum of eighteen squares," *Quart. J. Pure and Appl. Math. Oxford* **38** (1907), 289–351.
89. M.L. Glasser, private communication, April 1996.
90. M.L. Glasser and I.J. Zucker, *Lattice Sums*, Vol. 5 of Theoretical Chemistry: Advances and Perspectives (H. Eyring and D. Henderson, eds.), Academic Press, New York, 1980, pp. 67–139.
91. H.W. Gould, "Explicit formulas for Bernoulli numbers," *Amer. Math. Monthly* **79** (1972), 44–51.
92. I.P. Goulden and D.M. Jackson, *Combinatorial Enumeration*, John Wiley & Sons, New York, 1983.
93. I.S. Gradshteyn and I.M. Ryzhik, *Table of Integrals, Series, and Products*, 4th edn., Academic Press, San Diego, 1980; translated from the Russian by Scripta Technica, Inc., and edited by A. Jeffrey.
94. E. Grosswald, *Representations of Integers as Sums of Squares*, Springer-Verlag, New York, 1985.
95. K.-B. Gundlach, "On the representation of a number as a sum of squares," *Glasgow Math. J.* **19** (1978), 173–197.
96. R.A. Gustafson, "The Macdonald identities for affine root systems of classical type and hypergeometric series very well-poised on semi-simple Lie algebras," in *Ramanujan International Symposium on Analysis*, Pune, India, Dec. 26–28, 1987 (N.K. Thakare, ed.), 1989, pp. 187–224.
97. G.-N. Han, A. Randrianarivony, and J. Zeng, "Un autre q-analogue des nombres d'Euler," *Séminaire Lotharingien de Combinatoire* **42**(B42e) (1999), 22 pp.
98. G.-N. Han and J. Zeng, "q-Polynômes de Ghandi et statistique de Denert," *Discrete Math.* **205** (1999), 119–143.
99. G.H. Hardy, "On the representation of a number as the sum of any number of squares, and in particular of five or seven," *Proc. Nat. Acad. Sci., U.S.A.* **4** (1918), 189–193.
100. G.H. Hardy, "On the representation of a number as the sum of any number of squares, and in particular of five," *Trans. Amer. Math. Soc.* **21** (1920), 255–284.
101. G.H. Hardy, *Ramanujan*, Cambridge University Press, Cambridge 1940; reprinted by Chelsea , New York, 1978; reprinted by AMS Chelsea, Providence, RI, 1999; Now distributed by The American Mathematical Society, Providence, RI.
102. G.H. Hardy and E.M. Wright, *An Introduction to the Theory of Numbers*, 5th edn., Oxford University Press, Oxford, 1979.
103. J.B.H. Heilermann, "De transformatione serierum in fractiones continuas," Dr. Phil. Dissertation, Royal Academy of Münster, 1845.
104. J.B.H. Heilermann, "Über die Verwandlung der Reihen in Kettenbrüche," *J. Reine Angew. Math.* **33** (1846), 174–188.
105. H. Helfgott and I.M. Gessel, "Enumeration of tilings of diamonds and hexagons with defects," *Electron. J. Combin.* **6**(R16) (1999), 26 pp.
106. E. Hendriksen and H. Van Rossum, "Orthogonal moments," *Rocky Mountain J. Math.* **21** (1991), 319–330.
107. L.K. Hua, *Introduction to Number Theory*, Springer-Verlag, New York, 1982.
108. J.G. Huard, Z.M. Ou, B.K. Spearman, and K.S. Williams, "Elementary evaluation of certain convolution sums involving divisor functions," in *Number Theory for the Millennium* (M.A. Bennett, B.C. Berndt, N. Boston, H.G. Diamond, A.J. Hildebrand, and W. Philipp, eds.), Vol. 2, A.K. Peters, Natick, MA, to appear.
109. M.E.H. Ismail, J. Letessier, G. Valent, and J. Wimp, "Two families of associated Wilson polynomials," *Canad. J. Math.* **42** (1990), 659–695.
110. M.E.H. Ismail and D.R. Masson, "Generalized orthogonality and continued fractions," *J. Approx. Theory* **83** (1995), 1–40.
111. M.E.H. Ismail and D.R. Masson, "Some continued fractions related to elliptic functions," *Continued Fractions: From Analytic Number Theory to Constructive Approximation*, Columbia, MO, 1998 (B.C. Berndt and F. Gesztesy, eds.), Vol. 236 of Contemporary Mathematics, American Mathematical Society, Providence, RI, 1999, 149–166.

112. M.E.H. Ismail and M. Rahman, "The associated Askey-Wilson polynomials," *Trans. Amer. Math. Soc.* **328** (1991), 201–237.
113. M.E.H. Ismail and D. Stanton, "Classical orthogonal polynomials as moments," *Canad. J. Math.* **49** (1997), 520–542.
114. M.E.H. Ismail and D. Stanton, "More orthogonal polynomials as moments," *Mathematical Essays in Honor of Gian-Carlo Rota*, Cambridge, MA, 1996 (B.E. Sagan and R.P. Stanley, eds.), Vol. 161 of Progress in Mathematics, Birkhäuser Boston, Inc., Boston, MA, 1998, pp. 377–396.
115. M.E.H. Ismail and G. Valent, "On a family of orthogonal polynomials related to elliptic functions," *Illinois J. Math.* **42** (1998), 294–312.
116. M.E.H. Ismail, G. Valent, and G. Yoon, "Some orthogonal polynomials related to elliptic functions," *J. Approx. Theory* **112** (2001), 251–278.
117. C.G.J. Jacobi, "Fundamenta Nova Theoriae Functionum Ellipticarum," Regiomonti. Sumptibus fratrum Bornträger, 1829; reprinted in Jacobi's Gesammelte Werke, Vol. 1, Reimer, Berlin, 1881–1891, pp. 49–239; reprinted by Chelsea, New York, 1969; Now distributed by The American Mathematical Society, Providence, RI.
118. N. Jacobson, *Basic Algebra I*, W.H. Freeman and Co., San Francisco, CA, 1974.
119. W.B. Jones and W.J. Thron, "Continued Fractions: Analytic Theory and Applications," in *Encyclopedia of Mathematics and Its Applications*, Vol. 11 (G.-C. Rota, ed.), Addison-Wesley, London, 1980; Now distributed by Cambridge University Press, Cambridge.
120. V.G. Kac and M. Wakimoto, "Integrable highest weight modules over affine superalgebras and number theory," in *Lie Theory and Geometry*, in honor of Bertram Kostant (J.L. Brylinski, R. Brylinski, V. Guillemin and V. Kac, eds.), Vol. 123 of Progress in Mathematics, Birkhäuser Boston, Inc., Boston, MA, 1994, pp. 415–456.
121. V.G. Kac and M. Wakimoto, "Integrable highest weight modules over affine superalgebras and Appell's function," *Comm. Math. Phys.* **215** (2001), 631–682.
122. S. Karlin and G. Szegő, "On certain determinants whose elements are orthogonal polynomials," *J. Analyse Math.* **8** (1961), 1–157; reprinted in *Gabor Szegő: Collected Papers*, Vol. 3 (R. Askey, ed.), Birkhäuser Boston, Inc., Boston, MA, 1982, pp. 603–762.
123. M.I. Knopp, "On powers of the theta-function greater than the eighth," *Acta Arith.* **46** (1986), 271–283.
124. R. Koekoek and R.F. Swarttouw, "The Askey–scheme of hypergeometric orthogonal polynomials and its q-analogue," TU Delft, The Netherlands, 1998; available on the www: `ftp://ftp.twi.tudelft.nl/TWI/publications/ tech-reports/1998/DUT-TWI-98-17.ps.gz`.
125. C. Krattenthaler, "Advanced determinant calculus," *Séminaire Lotharingien de Combinatoire* **42**(B42q) (1999), 67 pp.
126. E. Krätzel, "Über die Anzahl der Darstellungen von natürlichen Zahlen als Summe von $4k$ Quadraten," *Wiss. Z. Friedrich-Schiller-Univ. Jena* **10** (1960/61), 33–37.
127. E. Krätzel, "Über die Anzahl der Darstellungen von natürlichen Zahlen als Summe von $4k+2$ Quadraten," *Wiss. Z. Friedrich-Schiller-Univ. Jena* **11** (1962), 115–120.
128. D.B. Lahiri, "On a type of series involving the partition function with applications to certain congruence relations," *Bull. Calcutta Math. Soc.* **38** (1946), 125–132.
129. D.B. Lahiri, "On Ramanujan's function $\tau(n)$ and the divisor function $\sigma_k(n)$—I," *Bull. Calcutta Math. Soc.* **38** (1946), 193–206.
130. D.B. Lahiri, "On Ramanujan's function $\tau(n)$ and the divisor function $\sigma_k(n)$—II," *Bull. Calcutta Math. Soc.* **39** (1947), 33–52.
131. D.B. Lahiri, "Identities connecting the partition, divisor and Ramanujan's functions," *Proc. Nat. Inst. Sci. India* **34A** (1968), 96–103.
132. D.B. Lahiri, "Some arithmetical identities for Ramanujan's and divisor functions," *Bull. Austral. Math. Soc.* **1** (1969), 307–314.
133. A. Lascoux, "Inversion des matrices de Hankel," *Linear Algebra Appl.* **129** (1990), 77–102.
134. D.F. Lawden, *Elliptic Functions and Applications*, Vol. 80 of Applied Mathematical Sciences, Springer-Verlag, New York, 1989.
135. B. Leclerc, "On identities satisfied by minors of a matrix," *Adv. in Math.* **100** (1993), 101–132.

136. B. Leclerc, "Powers of staircase Schur functions and symmetric analogues of Bessel polynomials," *Discrete Math.* **153** (1996), 213–227.
137. B. Leclerc, Private communication, July 1997.
138. B. Leclerc, "On certain formulas of Karlin and Szegő," *Séminaire Lotharingien de Combinatoire* **41**(B41d) (1998), 21 pp.
139. A.M. Legendre, "Traité des Fonctions Elliptiques et des Intégrales Euleriennes," t. III, Huzard-Courcier, Paris, 1828, pp. 133–134.
140. D.H. Lehmer, "Some functions of Ramanujan," *Math. Student* **27** (1959), 105–116.
141. J. Lepowsky, "Generalized Verma modules, loop space cohomology and Macdonald-type identities," *Ann. Sci. École Norm. Sup.* **12**(4) (1979), 169–234.
142. J. Lepowsky, "Affine Lie algebras and combinatorial identities," in *Lie Algebras and Related Topics*, Rutgers Univ. Press., New Brunswick, N.J., 1981, Vol. 933 of Lecture Notes in Math., Springer-Verlag, Berlin 1982, pp. 130–156.
143. G.M. Lilly and S.C. Milne, "The C_ℓ Bailey Transform and Bailey Lemma," *Constr. Approx.* **9** (1993), 473–500.
144. J. Liouville, "Extrait d'une lettre à M. Besge," *J. Math. Pures Appl.* **9**(2) (1864), 296–298.
145. D.E. Littlewood, *The Theory of Group Characters and Matrix Representations of Groups*, 2nd edn., Oxford University Press, Oxford, 1958.
146. Z.-G. Liu, "On the representation of integers as sums of squares," in *q-Series with Applications to Combinatorics, Number Theory, and Physics* (B.C. Berndt and Ken Ono, eds.), Vol. 291 of Contemporary Mathematics, American Mathematical Society, Providence, RI, 2001, pp. 163–176.
147. G.A. Lomadze, "On the representation of numbers by sums of squares," *Akad. Nauk Gruzin. SSR Trudy Tbiliss. Mat. Inst. Razmadze* **16** (1948), 231–275. (in Russian; Georgian summary).
148. J.S. Lomont and J.D. Brillhart, *Elliptic Polynomials*, Chapman & Hall/CRC Press, Boca Raton, FL, 2000.
149. L. Lorentzen and H. Waadeland, *Continued Fractions With Applications*, Vol. 3 of Studies in Computational Mathematics, North-Holland, Amsterdam, 1992.
150. J. Lützen, "Joseph Liouville 1809–1882: Master of pure and applied mathematics," in *Studies in the History of Mathematics and Physical Sciences*, Vol. 15, Springer-Verlag, New York, 1990.
151. I.G. Macdonald, "Affine root systems and Dedekind's η-function," *Invent. Math.* **15** (1972), 91–143.
152. I.G. Macdonald, "Some conjectures for root systems," *SIAM J. Math. Anal.* **13** (1982), 988–1007.
153. I.G. Macdonald, *Symmetric Functions and Hall Polynomials*, 2nd edn., Oxford University Press, Oxford, 1995.
154. G.B. Mathews, "On the representation of a number as a sum of squares," *Proc. London Math. Soc.* **27** (1895–96), 55–60.
155. H. McKean and V. Moll, *Elliptic Curves: Function Theory, Geometry, Arithmetic*, Cambridge University Press, Cambridge, 1997.
156. M.L. Mehta, *Elements of Matrix Theory*, Hindustan Publishing Corp., Delhi, 1977.
157. M.L. Mehta, "Matrix theory: Selected topics and useful results," *Les Editions de Physique*, Les Ulis, France, 1989; see Appendix A.5 (In India, sold and distributed by Hindustan Publishing Corp.).
158. S.C. Milne, "An elementary proof of the Macdonald identities for $A_\ell^{(1)}$," *Adv. in Math.* **57** (1985), 34–70.
159. S.C. Milne, "Basic hypergeometric series very well-poised in $U(n)$," *J. Math. Anal. Appl.* **122** (1987), 223–256.
160. S.C. Milne, "Classical partition functions and the $U(n+1)$ Rogers-Selberg identity," *Discrete Math.* **99** (1992), 199–246.
161. S.C. Milne, "The C_ℓ Rogers-Selberg identity," *SIAM J. Math. Anal.* **25** (1994), 571–595.
162. S.C. Milne, "New infinite families of exact sums of squares formulas, Jacobi elliptic functions, and Ramanujan's tau function," *Proc. Nat. Acad. Sci., U.S.A.* **93** (1996), 15004–15008.
163. S.C. Milne, "Balanced ${}_3\phi_2$ summation theorems for $U(n)$ basic hypergeometric series," *Adv. in Math.* **131** (1997), 93–187.
164. S.C. Milne, "Hankel determinants of Eisenstein series," in *Symbolic Computation, Number Theory, Special Functions, Physics and Combinatorics*, Gainesville, 1999 (F.G. Garvan and M. Ismail, eds.), Vol. 4 of Dev. Math., Kluwer Academic Publishers, Dordrecht, 2001, pp. 171–188.
165. S.C. Milne, "A new formula for Ramanujan's tau function and the Leech lattice," in preparation.

166. S.C. Milne, "Continued fractions, Hankel determinants, and further identities for powers of classical theta functions," in preparation.
167. S.C. Milne, "Sums of squares, Schur functions, and multiple basic hypergeometric series," in preparation.
168. S.C. Milne and G.M. Lilly, "The A_ℓ and C_ℓ Bailey transform and lemma," *Bull. Amer. Math. Soc. (N.S.)* **26** (1992), 258–263.
169. S.C. Milne and G.M. Lilly, "Consequences of the A_ℓ and C_ℓ Bailey transform and Bailey lemma," *Discrete Math.* **139** (1995), 319–346.
170. S.C. Mitra, "On the expansion of the Weierstrassian and Jacobian elliptic functions in powers of the argument," *Bull. Calcutta Math. Soc.* **17** (1926), 159–172.
171. L.J. Mordell, "On Mr. Ramanujan's empirical expansions of modular functions," *Proc. Cambridge Philos. Soc.* **19** (1917), 117–124.
172. L.J. Mordell, "On the representation of numbers as the sum of $2r$ squares," *Quart. J. Pure and Appl. Math. Oxford* **48** (1917), 93–104.
173. L.J. Mordell, "On the representations of a number as a sum of an odd number of squares," *Trans. Cambridge Philos. Soc.* **22** (1919), 361–372.
174. T. Muir, "New general formulae for the transformation of infinite series into continued fractions," *Trans. Roy. Soc. Edinburgh* **27** (1872–1876; see Part IV. 1875–76), 467–471.
175. T. Muir, "On the transformation of Gauss' hypergeometric series into a continued fraction," *Proc. London Math. Soc.* **7** (1876), 112–119.
176. T. Muir, "On Eisenstein's continued fractions," *Trans. Roy. Soc. Edinburgh* **28** (1876–1878; see Part I. 1876–1877), 135–143.
177. T. Muir, *The Theory of Determinants in the Historical Order of Development*, Vol. I (1906), Vol. II (1911), Vol. III (1920), Vol. IV (1923), Macmillan and Co., Ltd., London.
178. T. Muir, "The theory of persymmetric determinants in the historical order of development up to 1860," *Proc. Roy. Soc. Edinburgh* **30** (1910), 407–431.
179. T. Muir, "The theory of persymmetric determinants from 1894 to 1919," *Proc. Roy. Soc. Edinburgh* **47** (1926–27), 11–33.
180. T. Muir, *Contributions to the History of Determinants 1900–1920*, Blackie & Son, London and Glasgow, 1930.
181. T. Muir, *A Treatise on the Theory of Determinants*, Dover Publications, New York, 1960.
182. M.B. Nathanson, *Elementary Methods in Number Theory*, Vol. 195 of Graduate Texts in Mathematics, Springer-Verlag, New York, 2000.
183. K. Ono, "Representations of integers as sums of squares," *J. Number Theory*, to appear.
184. K. Ono, S. Robins, and P.T. Wahl, "On the representation of integers as sums of triangular numbers," *Aequationes Math.* **50** (1995), 73–94.
185. O. Perron, *Die Lehre von den Kettenbrüchen*, 2nd edn., B.G. Teubner, Leipzig and Berlin, 1929; reprinted by Chelsea, New York, 1950.
186. Von K. Petr, "Über die Anzahl der Darstellungen einer Zahl als Summe von zehn und zwölf Quadraten," *Archiv Math. Phys.* **11**(3) (1907), 83–85.
187. B. van der Pol, "The representation of numbers as sums of eight, sixteen and twenty-four squares," *Nederl. Akad. Wetensch. Proc. Ser. A* **57** (1954), 349–361; *Nederl. Akad. Wetensch. Indag. Math.* **16** (1954), 349–361.
188. G. Prasad, *An Introduction to the Theory of Elliptic Functions and Higher Transcendentals*, University of Calcutta, 1928.
189. H. Rademacher, *Topics in Analytic Number Theory*, Vol. 169 of Grundlehren Math. Wiss., Springer-Verlag, New York, 1973.
190. Ch. Radoux, "Calcul effectif de certains déterminants de Hankel," *Bull. Soc. Math. Belg. Sér* B **31** (1979), 49–55.
191. Ch. Radoux, "Déterminant de Hankel construit sur les polynômes de Hérmite," *Ann. Soc. Sci. Bruxelles Sér. I* **104** (1990), 59–61.
192. Ch. Radoux, "Déterminant de Hankel construit sur des polynômes liés aux nombres de dérangements," *European J. Combin.* **12** (1991), 327–329.
193. Ch. Radoux, "Déterminants de Hankel et théorème de Sylvester," Actes de la 28e session du Séminaire Lotharingien de Combinatoire, publication de l'I.R.M.A. No. 498/S–28, Strasbourg, 1992, pp. 115–122.

194. S. Ramanujan, "On certain arithmetical functions," *Trans. Cambridge Philos. Soc.* **22** (1916), 159–184; reprinted in *Collected Papers of Srinivasa Ramanujan*, Chelsea, New York, 1962, pp. 136–162; reprinted by AMS Chelsea, Providence, RI, 2000; Now distributed by The American Mathematical Society, Providence, RI.
195. S. Ramanujan, *The Lost Notebook and Other Unpublished Papers*, Narosa, New Delhi, 1988.
196. A. Randrianarivony, "Fractions continues, combinatoire et extensions de nombres classiques," Ph.D. Thesis, Univ. Louis Pasteur, Strasbourg, France, 1994.
197. A. Randrianarivony, "Fractions continues, q-nombres de Catalan et q-polynômes de Genocchi," *European J. Combin.* **18** (1997), 75–92.
198. A. Randrianarivony, "q, p-analogue des nombres de Catalan," *Discrete Math.* **178** (1998), 199–211.
199. A. Randrianarivony and J.A. Zeng, "Extension of Euler numbers and records of up-down permutations," *J. Combin. Theory Ser.* A **68** (1994), 86–99.
200. A. Randrianarivony and J. Zeng, "A family of polynomials interpolating several classical series of numbers," *Adv. in Appl. Math.* **17** (1996), 1–26.
201. R.A. Rankin, "On the representations of a number as a sum of squares and certain related identities," *Proc. Cambridge Philos. Soc.* **41** (1945), 1–11.
202. R.A. Rankin, "On the representation of a number as the sum of any number of squares, and in particular of twenty," *Acta Arith.* **7** (1962), 399–407.
203. R.A. Rankin, "Sums of squares and cusp forms," *Amer. J. Math.* **87** (1965), 857–860.
204. R.A. Rankin, *Modular Forms and Functions*, Cambridge University Press, Cambridge, 1977.
205. D. Redmond, *Number Theory: An Introduction*, Marcel Dekker, New York, 1996.
206. D.P. Robbins, "Solution to problem **10387***," *Amer. Math. Monthly* **104** (1997), 366–367.
207. L.J. Rogers, "On the representation of certain asymptotic series as convergent continued fractions," *Proc. London Math. Soc* **4**(2) (1907), 72–89.
208. A. Schett, "Properties of the Taylor series expansion coefficients of the Jacobian elliptic functions," *Math. Comp.* **30** (1976), 143–147, with microfiche supplement (See also: "Corrigendum," *Math. Comp.* **31** (1977), 330).
209. A. Schett, "Recurrence formula of the Taylor series expansion coefficients of the Jacobian elliptic functions," *Math. Comp.* **31** (1977), 1003–1005, with microfiche supplement.
210. W. Seidel, "Note on a persymmetric determinant," *Quart. J. Math., Oxford Ser.* **4**(2) (1953), 150–151.
211. J.-P. Serre, *A Course in Arithmetic*, Vol. 7 of Graduate Texts in Mathematics, Springer-Verlag, New York, 1973.
212. W. Sierpinski, "Wzór analityczny na pewna funkcje liczbowa (Une formule analytique pour une fonction numérique)," *Wiadomości Matematyczne Warszawa* **11** (1907), 225–231 (in Polish).
213. H.J.S. Smith, *Report on the Theory of Numbers*, Part VI (Report of the British Association for 1865, pp. 322–375), 1894; reprinted in *The Collected Mathematical Papers of H.J.S. Smith*, Vol. 1 (J.W.L. Glaisher, ed.), 1894, pp. 306–311; reprinted by Chelsea, New York, 1965.
214. H.J.S. Smith, "On the orders and genera of quadratic forms containing more than 3 indeterminates," *Proc. Roy. Soc. London* **16** (1867), 197–208; reprinted in *The Collected Mathematical Papers of H.J.S. Smith*, Vol. 1 (J.W.L. Glaisher, ed.), 1894, pp. 510–523; reprinted by Chelsea, New York, 1965.
215. R.P. Stanley, *Enumerative Combinatorics*, Vol. I, Wadsworth & Brooks Cole, Belmont, CA, 1986.
216. M.A. Stern, "Theorie der Kettenbrüche und ihre Anwendung," *J. Reine Angew. Math.* **10** (1833), 1–22, 154–166, 241–274, 364–376.
217. M.A. Stern, "Theorie der Kettenbrüche und ihre Anwendung," *J. Reine Angew. Math.* **11** (1834), 33–66, 142–168, 277–306, 311–350.
218. T.J. Stieltjes, "Sur la réduction en fraction continue d'une série procédant suivant les puissances descendantes d'une variable," *Ann. Fac. Sci. Toulouse* **3** (1889), H. 1–17; reprinted in *Œuvres Complètes* **T2**, P. Noordhoff, Groningen, 1918, pp. 184–200; see also *Œuvres Complètes (Collected Papers)*, Vol. II (G. van Dijk, ed.), Springer-Verlag, Berlin, 1993, pp. 188–204.
219. T.J. Stieltjes, "Sur quelques intégrales définies et leur développement en fractions continues," *Quart. J. Math.* **24** (1890), 370–382; reprinted in *Œuvres Complètes* **T2**, P. Noordhoff, Groningen, 1918, pp. 378–391; see also *Œuvres Complètes (Collected Papers)* Vol. II (G. van Dijk, ed.), Springer-Verlag, Berlin, 1993, pp. 382–395.

220. T.J. Stieltjes, "Recherches sur les fractions continues," *Ann. Fac. Sci. Toulouse* **8** (1894), J. 1–122, **9** (1895), A. 1–47; reprinted in *Œuvres Complètes* **T2**, P. Noordhoff, Groningen, 1918, pp. 402–566; (see pp. 549–554); see also *Œuvres Complètes (Collected Papers)*, Vol. II (G. van Dijk, ed.), Springer-Verlag, Berlin, 1993, pp. 406–570 (see also pp. 609–745 for an English translation. Note especially pp. 728–733).

221. O. Szász, "Über Hermitesche Formen mit rekurrierender Determinante und über rationale Polynome," *Math. Z.* **11** (1921), 24–57.

222. G. Szegő, "On an inequality of Turán concerning Legendre polynomials," *Bull. Amer. Math. Soc.* **54** (1948), 401–405; reprinted in *Gabor Szegő: Collected Papers*, Vol. 3, 1945–1972 (R. Askey, ed.), Birkhäuser Boston, Inc., Boston, MA, 1982, pp. 69–73, 74–75.

223. O. Taussky, "Sums of squares," *Amer. Math. Monthly* **77** (1970), 805–830.

224. J. Touchard, "Sur un probléme de configurations et sur les fractions continues," *Canad. J. Math.* **4** (1952), 2–25.

225. P. Turán, "On the zeros of the polynomials of Legendre," *Časopis pro Pěstováni Matematiky a Fysiky* **75** (1950), 113–122.

226. H.W. Turnbull, *The Theory of Determinants, Matrices, and Invariants*, Blackie and Son, London, 1928; reprinted by Dover Publications, New York, 1960.

227. J.V. Uspensky, "Sur la représentation des nombres par les sommes des carrés," *Communications de la Société mathématique de Kharkow série 2* **14** (1913), 31–64 (in Russian).

228. J.V. Uspensky, "Note sur le nombre de représentations des nombres par une somme d'un nombre pair de carrés," *Bulletin de l'Académie des Sciences de l'URSS, Leningrad (Izvestija Akademii Nauk Sojuza Sovetskich Respublik. Leningrad.) Serie 6* **19** (1925), 647–662 (in French).

229. J.V. Uspensky, "On Jacobi's arithmetical theorems concerning the simultaneous representation of numbers by two different quadratic forms," *Trans. Amer. Math. Soc.* **30** (1928), 385–404.

230. J.V. Uspensky and M.A. Heaslet, *Elementary Number Theory*, McGraw-Hill, New York, 1939.

231. G. Valent, "Asymptotic analysis of some associated orthogonal polynomials connected with elliptic functions," *SIAM J. Math. Anal.* **25** (1994), 749–775.

232. G. Valent, "Associated Stieltjes-Carlitz polynomials and a generalization of Heun's differential equation," *J. Comput. Appl. Math.* **57** (1995), 293–307.

233. G. Valent and W. Van Assche, "The impact of Stieltjes' work on continued fractions and orthogonal polynomials: Additional material," *J. Comput. Appl. Math.* **65** (1995), 419–447; this volume was devoted to the Proceedings of the International Conference on Orthogonality, Moment Problems and Continued Fractions (Delft, 1994).

234. W. Van Assche, "Asymptotics for orthogonal polynomials and three-term recurrences," in *Orthogonal Polynomials: Theory and Practice* (P. Nevai, ed.), Vol. 294 of NATO-ASI Series C: Mathematical and Physical Sciences, Kluwer Academic Publishers, Dordrecht, 1990, pp. 435–462.

235. W. Van Assche, "The impact of Stieltjes work on continued fractions and orthogonal polynomials," Vol. I of *T.J. Stieltjes: Œuvres Complètes (Collected Papers)* (G. van Dijk, ed.), Springer-Verlag, Berlin, 1993, pp. 5–37.

236. P.R. Vein, "Persymmetric determinants. I. The derivatives of determinants with Appell function elements," *Linear and Multilinear Algebra* **11** (1982), 253–265.

237. P.R. Vein, "Persymmetric determinants. II. Families of distinct submatrices with nondistinct determinants," *Linear and Multilinear Algebra* **11** (1982), 267–276.

238. P.R. Vein, "Persymmetric determinants. III. A basic determinant," *Linear and Multilinear Algebra* **11** (1982), 305–315.

239. P.R. Vein, "Persymmetric determinants. IV. An alternative form of the Yamazaki-Hori determinantal solution of the Ernst equation," *Linear and Multilinear Algebra* **12** (1982/83), 329–339.

240. P.R. Vein, "Persymmetric determinants. V. Families of overlapping coaxial equivalent determinants," *Linear and Multilinear Algebra* **14** (1983), 131–141.

241. P.R. Vein and P. Dale, "Determinants, their derivatives and nonlinear differential equations," *J. Math. Anal. Appl.* **74** (1980), 599–634.

242. B.A. Venkov, *Elementary Number Theory*, Wolters-Noordhoff Publishing, Groningen, 1970; translated from the Russian and edited by Helen Alderson (Popova).

243. R. Vermes, "Hankel determinants formed from successive derivatives," *Duke Math. J.* **37** (1970), 255–259.

244. G. Viennot, "Une interprétation combinatoire des coefficients des développements en série entière des fonctions elliptiques de Jacobi," *J. Combin. Theory Ser. A* **29** (1980), 121–133.
245. G. Viennot, "Une théorie combinatoire des polynômes orthogonaux généraux," in *Lecture Notes*, publication de l'UQAM, Montréal (1983).
246. G. Viennot, "A combinatorial interpretation of the quotient-difference algorithm," Technical Report No. 8611, Université de Bordeaux I, 1986.
247. A.Z. Walfisz, "On the representation of numbers by sums of squares: Asymptotic formulas," *Uspehi Mat. Nauk (N.S.)* **52**(6) (1952), 91–178 (in Russian); English transl. in *Amer. Math. Soc. Transl.* **3**(2) (1956), 163–248.
248. H.S. Wall, *Analytic Theory of Continued Fractions*, D. Van Nostrand, New York, 1948; reprinted by Chelsea, New York, 1973.
249. J.B. Walton, "Theta series in the Gaussian field," *Duke Math. J.* **16** (1949), 479–491.
250. E.T. Whittaker and G.N. Watson, *A Course of Modern Analysis*, 4th edn., Cambridge University Press, Cambridge, 1927.
251. S. Wolfram, *The Mathematica Book*, 4th edn., Wolfram Media/Cambridge University Press, Cambridge, 1999.
252. S. Wrigge, "Calculation of the Taylor series expansion coefficients of the Jacobian elliptic function $sn(x, k)$," *Math. Comp.* **36** (1981), 555–564.
253. S. Wrigge, "A note on the Taylor series expansion coefficients of the Jacobian elliptic function $sn(x, k)$," *Math. Comp.* **37** (1981), 495–497.
254. D. Zagier, "A proof of the Kac-Wakimoto affine denominator formula for the strange series," *Math. Res. Letters* **7** (2000), 597–604.
255. D. Zeilberger, "Proof of the alternating sign matrix conjecture," *Electron. J. Combin.* **3**(R13) (1996), 84 pp.
256. D. Zeilberger, "Proof of the refined alternating sign matrix conjecture," *New York J. Math.* **2** (1996), 59–68.
257. J. Zeng, "Énumérations de permutations et J-fractions Continues," *European J. Combin.* **14** (1993), 373–382.
258. J. Zeng, "Sur quelques propriétes de symétrie des nombres de Genocchi," *Discrete Math.* **153** (1996), 319–333.
259. I. J. Zucker, "The summation of series of hyperbolic functions," *SIAM J. Math. Anal.* **10** (1979), 192–206.

